T0238483

BestMasters

Sebastian Uhlig

Self-Organized Surface Structures with Ultrafast White-Light

First Investigation of LIPSS with Supercontinuum

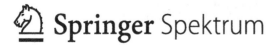 Springer Spektrum

Sebastian Uhlig
Fraunhofer Institute for Photonic
  Microsystems
Dresden, Germany

BestMasters
ISBN 978-3-658-09893-3   ISBN 978-3-658-09894-0 (eBook)
DOI 10.1007/978-3-658-09894-0

Library of Congress Control Number: 2015938767

Springer Spektrum

Printed on acid-free paper

Springer Spektrum is a brand of Springer Fachmedien Wiesbaden
Springer Fachmedien Wiesbaden is part of Springer Science+Business Media
(www.springer.com)

Table of Contents

Abbrevations, Constants & Symbols

p.a.m.f.	picture adapted and modified from
ns	Nanoseconds
ps	Picoseconds
fs	Femtoseconds
AFM	Atomic Force Microscope
CW	Continuous Wave
EDX	Energy Dispersive X-ray Spectroscopy
FWHM	Full Width Half Maximum
IBS	Ion Beam Sputtering
IR	Infrared
LIPSS	Laser-Induced Periodic Surface Structure
MPE	Multi-Photon Excitation
MW	Mega-Watt
NDF	Neutral Density Filters
SEM	Scanning Electron Microscope
SCG	Supercontinuum Generation
SF	Self-Focusing
SPF	Short-Pass Filter
SPM	Self-Phase Modulation
SPP	Surface Plasmon Polaritons
UV	Ultra-Violet
VIS	Visible
WL	White-Light
WLC	White-Light Continuum

Constants

Symbol	Quantity	Value
ε_0	Vacuum Permittivity	$= 8.8541878 \times 10^{-12}$ F/m
c	Speed of Light in Vacuum	$= 299792458$ m/s
k_B	Boltzman Constant	$= 1.3806488 \times 10^{-23}$ J/K
m_e	Electron Mass	$= 9.10938212 \times 10^{-31}$ kg
e	Electron Charge	$= 1.602176487 \times 10^{-19}$ C
n_0	Al_2O_3 Refractive Index	$= 1.76$ (@ 800 nm)

Frequently used Symbols

Symbol	Quantity
α	Rotation Angle of Wave-plate Retarder / Analyzer
E_p	Laser Pulse Energy
E_0	Laser Pulse Energy at the Position of the Crystal
E_d	Deconvolved White-Light Pulse Energy & Filtered WL Pulse Energy
E_{max}	Maximum Possible Pump Pulse Energy
E_{wl}	Threshold Pump Pulse Energy at Onset of WL Generation
$\vec{E}(r,t)$	Electric Field Vector
f_{rep}	Laser Pulse Repetition Frequency
F_{abl}	Threshold Fluence for Single Pulse Ablation
I	Intensity
I_D	Irradiation Dose
h	Corrugation Height
λ_{laser}	Central Laser Wavelength
λ_{wl}	Wavelength within White-Light Spectrum
λ_{c_i}	Spectral Cut-Off Wavelength
Λ_{LIPSS}	Structure Period Length
Λ_{pri}	Spatial Period Length of Primary Pattern
Λ_{sec}	Spatial Period Length of Secondary Pattern
Λ_{fine}	Spatial Period Length of Fine Pattern
n	Refractive Index
n_2	Nonlinear Refractive Index
$\vec{P}(r,t)$	Induced Polarization of a Medium
$\vec{k}, k$	Wavevector, Wavenumber
σ	Standard Deviation
θ	Angle of Incident
V	Visibility
$w_{0,i}$	Gaussian Beam Waist Size in Transverse Direction
ω	Electric Field Frequency
$\underline{\chi}^{(1)}(\omega)$	Linear Susceptibility Tensor
χ_{eff}	Effective Susceptibility
χ_{NL}	Nonlinear Contribution to Susceptibility

Abstract

This Master Thesis presents the first experimental investigation of self-organized surface structures (LIPSS) generated by ablation from different (semiconductor and metallic) targets with an ultrafast white-light continuum (WLC) spreading in wavelength from 400-750 nm.

The main goal of this work is to study the possibility of LIPSS formation upon irradiation with an incoherent and polychromatic light source (e.g. the WLC), in order to discriminate between the two debated formation scenarios. The generation of a suitable WLC in terms of sufficient white-light pulse energy, broad spectral bandwidth and low spatial coherence for the LIPSS generation, as well as the characterization of this WLC are furthermore important objectives.

By pumping an Al_2O_3 crystal with ultrafast (800 nm, 75 fs, 1 kHz, 270 μJ) laser pulses, a resulting continuum spreading from $\sim 400 - 1000$ nm is obtained. It is subsequently filtered for the visible part of the spectrum ($\sim 400 - 750$ nm), blocking out the infrared part and highly coherent pump-laser. The characteristics of the white-light continuum are investigated, with a focus on energy conversion efficiency, white-light polarization, spatial beam properties, long term stability and spatial coherence. It has been established that the spatial coherence of the white-light beam at high pump energy is significantly reduced in comparison to the pump irradiation. About 14% of the pump energy is converted to wavelengths below < 750 nm. The white-light beam exhibits a spatial dependence of the wavelengths distribution and preserves the linear polarization of the pump laser.

Typical LIPSS structures were produced on silicon (with periods between 600...700 nm), brass (~ 500 nm), copper ($\sim 300...500$ nm) and stainless steel ($\sim 300...450$ nm) upon multi-pulse irradiation from the white-light continuum. The patterns consist of long bifurcating lines, where the orientation is influenced through polarization of the white-light light. The ripples periods depend, clearly, on both, the material and the irradiation dose (fluence $\times$ number of pulses) with an exponentially decreasing behavior, upon increasing dose, on silicon and stainless steel. Additionally, the dependence of the spatial periods of LIPSS on the local intensity of the white-light beam, as well as a coexistence of fine and coarse ripples and evolution of the surface patterns upon multi-pulse irradiation was observed.

The experimental results indicate that it is not necessary to have a coherent light source for the generation of periodic surface patterns on solid targets. Given the continuous excitation spectrum with very moderate power in narrow spectral intervals, it appears unlikely to attribute the structure formation to any interference effect. Instead, the results are in full agreement with the *dynamic* model of self-organized structure formation.

Introduction

Laser induced periodic surface structures (LIPSS) often termed "ripples", have been an active field of research ever since their discovery in 1965 [Bir65], shortly after the realization of the first laser in 1960 [Mai60].

The early investigation of ripple patterns with nanosecond laser pulses, showed a periodicity depending on the laser wavelengths and on the angle of incidence. Moreover, most patterns exhibit an orientation perpendicular to the laser polarization. Thus, an interference effect of the incident laser and a secondary surface electromagnetic wave (e.g. generated by scattering of laser irradiation) was suggested, as the fundamental physical process involved in the formation of these structures [SYP+83] (often referred to as "classical ripple theory").

Through the investigation of LIPSS with femtosecond laser pulses a variety of other structures as well as ripple patterns with spatial periods well below the incident laser wavelengths were observed. The classical ripple theory has therefore been expanded to the, in this thesis referred to as *static* model of structure formation [BRK09], [GCP+11]. The so called "femtosecond ripples" share many common characteristics in the morphological appearance (bifurcation and truncations of lines, dot & columnar structures) with self-organized pattens from induced surface instability after ion beam bombardment [ZCF+09]. Furthermore, the structure sizes produced upon femtosecond laser irradiation or ion beam sputtering, seem to depend on the laser intensity/ ion energy and irradiation time as well as on a positive feedback, which results in the coalescence of structures. This leads to the suggestion that other physical process may be involved in the formation process. The adaptation of the theory of ion beam sputtering [Sig69], [BHa88], [Var13b] gave rise to the *dynamic* model of self-organized LIPSS formation.

Despite numerous experimental data on surface patterning upon femtosecond laser irradiation, the microscopic mechanisms of LIPSS formation are still under discussion. Supporting and contrary arguments for the two formation models, can be found in the observed features of the LIPSS patterns. For instance, grating-like structures with spatial periods in the magnitude of the incident wavelength ($\Lambda_{\text{LIPSS}} \sim \lambda_{laser}$), may result from an inhomogeneous ablation through the interference pattern of the interacting waves. Thereby, the monochromaticity and high spatial coherence of the laser, are essential properties for inference and, thus, formation of periodic ripple patterns according to the *static* model.

On the other hand, spherical nano-dots upon irradiation with circular polarized laser light, bifurcating ripples, and patterns with periods larger or significantly smaller than the laser wavelength are features which cannot be accounted for by an optical interference effect. They can, however, be explained in the framework of a self-organization process from laser-induced instability [Cos06], [Rei10], [Var13b].

To shed light on the controversy between these two formation models, the LIPSS formation ought to be investigated with a polychromatic light source of reduced spatial coherence. As a result of this, an optical interference effect might be excluded or confirmed as the driving physical process in LIPSS formation.

This open question has been the major motivation for this master thesis. Spontaneous pattern formation upon irradiation with incoherent white-light from an incandescent light bulb, has already been reported in photo-refractive crystals [TBS04], [TTB+04], [KSE+00]. However, to study the features of the "femtosecond ripples", the intended light source further needs to provide intense pulses in the ultrafast time regime, with duration below one picosecond ($\tau_p \leq 1\text{ps}$).

Since 1970, it has been known that narrow bandwidth laser pulses can undergo an extreme spectral broadening upon propagation through a nonlinear optical medium [ASh70]. The so called "generation of a white-light supercontinuum" is nowadays an active field of research, since the applications of the white-light are very attractive to many scientific fields, e.g. optical tomography [HLC+01]. The resulting white-light beam, thereby, exhibits very distinct characteristics in terms of spatial wavelengths dependence as well as spatial beam properties at varying input laser powers. Most importantly the white-light pulses stay ultrafast light pulses upon propagation through a nonlinear optical medium of short length [QWW+01], [NOS+96].

The investigation of the spatial coherence of the white-light continuum shows two different sides, depending of the input power of the pumping laser. It has been published that, the white-light retains the high spatial coherence of the pump-laser [WIt01], [CPB+99] for low input powers, making it a polychromatic laser source. However, for large input powers, it has been reported [BIC96], [GSE86] that the white-light looses its spatial coherence.

For these reasons, it has been a motivation, to characterize the generated white-light continuum, previous to the study of the LIPSS formation.

The goal of this work is to study the possibility of LIPSS formation upon irradiation with a suitable white-light continuum of sufficient WL pulse energy, broad spectral bandwidth and low spatial coherence. Furthermore, to discriminate between both LIPSS formation models by experimentally studying the importance of laser wavelength and coherence vs. irradiation dose on the characteristic features of LIPSS.

Thesis Organization

The first two chapters of this thesis are intended to give a brief theoretical introduction into the two major scientific fields of laser ablation & laser surface structuring and nonlinear optics. Presented in **chapter one**, are the favored mechanism of femtosecond laser ablation on metals and semiconductors as well as the two formation models of LIPSS. The, in this work, preferred terminology of the structures and spatial period lengths is also given here.

In **chapter two**, the nonlinear optical processes responsible for the spectral broadening of laser pulses (i.e. continuum generation) with a focus on the used nonlinear optical medium Sapphire is discussed.

The experimental setup for the characterization of the white-light continuum and studies of LIPSS is presented in **chapter three**.

Chapter four discusses the experimental results from the characterization of the white-light continuum with a focus on energy conversion efficiency, white-light polarization, spatial beam properties, long term stability and spatial coherence.

In **chapter five** the study of LIPSS upon irradiation with the white-light continuum is presented. The investigation of the produced structures focuses on the multi-pulse feedback effect as well as on the identification and characterization of similarities in the observed structures to self-organized surface patterns upon laser irradiation or ion beam sputtering. Furthermore, the effect of the white-light polarization on the orientation of the ripple patterns as well as energy dispersive x-ray measurements of the surface element distribution after irradiation in air atmosphere are discussed.

The conclusion of this work and outlook on future experiments is given in **chapter six**

In addition, the determination and detailed description of the pulse duration and intensity distribution in temporal and spatial domain, as well as the beam propagation of the laser beam that was used in the experiments of this work are presented in the **appendix.**

Publication, Presentations and Posters to this Work

1. J. Reif, O. Varlamova, S. Uhlig, S.Varlamov, M. Bestehorn. Formation of self-organized LIPSS by irradiation with an ultrafast White-Light Continuum. *Proceedings of LPM 2014 - 15th international Symposium on laser precision microfabrication*

2. S. Uhlig, O. Varlamova, M. Ratzke, J. Reif. Formation of self-organized LIPSS by irradiation with an ultrafast White-Light Continuum. *Poster at EMRS Spring Meeting 2014*

3. J. Reif, O. Varlamova, S. Uhlig, S. Varlamov, M. Bestehorn. On the Physics of Self-Organized Nanostructure Formation upon Femtosecond Laser Ablation. *Applied Physics A,114(4):1-6,2014*, ISSN 0947-8396, http://dx.doi.org/10.1007/s00339-014-8339-x

4. J. Reif, O. Varlamova, S. Uhlig, C. Martens. Genesis of Femtosecond-Induced Nanostructures on Solid Surfaces. *Presentation at 3rd workshop on LIPSS (2nd circular), Berlin November 7th 2013*

5. J. Reif, O. Varlamova, S. Uhlig, S.Varlamov, M. Bestehorn. On the Physics of Self-Organized Nanostructure Formation upon Femtosecond Laser Ablation. *Proceedings of LMP 2013 - 14th International Symposium on Laser Precision Microfabrication*

Chapter 1

Introduction to Laser-Ablation &-Surface Structuring

1.1 Femtosecond Laser Ablation of Metals and Semiconductors

Femtosecond (fs) laser ablation is the rapid removal of material from the surface region of a target, induced by laser pulses with duration in the order of tens to hundreds of femtoseconds (10^{-15}s). Thereby the duration of the laser pulse is so short that it addresses only the target electrons. Every other process and response of the material takes place after the pulse.

In semiconductors, absorption by a valence band electron results in the creation of an electron-hole pair in the band structure. For silicon, as an indirect semiconductor, this can be achieved by linear absorption of one photon and help of a phonon (for momentum conversation), or via nonlinear mechanism by simultaneous absorption of multiple photons. In metals, the absorption happens by the conduction electrons via inverse bremsstrahlung. The electrons thereby gain higher kinetic energy (i.e. heat). Highly excited electrons can be emitted, during the duration of the pulse, from the near-surface region.

The favored mechanism of ablation for metals and semiconductors by intense ∼100fs laser pulses involves a non-thermal melting process of the perturbed surface region [SCS99], [MKe95].

In Fig.(1.1) a molecular dynamics simulation of an ablation sequence in silicon is displayed [HKF⁺04]. Here, physical system is represented by a number of particles. The system is irradiated by 100fs laser pulse with Gaussian intensity distribution. The simulation starts at $t = 0.0$ps and the laser pulse arrives at $t = 1$ps.

Within hundreds of fs after the pulse the system is highly perturbed, however the crystalline structure still reflects the equilibrium state, which means the average displacement from the equilibrium lattice is still zero [LLS⁺05]. The state of matter at this point is an intermediate between a solid and liquid.

The excited electrons transfer their kinetic energy to the initially cold lattice via electron-phonon collisions. Within tens of picoseconds ($t < 10$ps) the system heats up and exceeds the melting temperature. Now a highly pressurized (∼ 10GPa) and superheated non-equilibrium liquid layer [LLM03] forms in the affected surface region.

The number of nearest neighbors of a particle becomes larger, because solid silicon does not exhibit the highest density sphere package. As a result the bandgap disappears, due to overlapping energy state and silicon, like other semiconductors, is metallic in the liquid state (in Fig. (1.1) displayed as coordination number CN¿4, (solid silicon:CN=4). On such a timescale the system has no time for a normal boiling process, which usually takes t>100ns and is in thermal equilibrium. While the melt front propagates further into the bulk with supersonic speed, more solid material is transferred to the liquid phase. The unstable region relaxes through a rapid mechanical expansion into vacuum via *phase explosion* (explosive boiling) [MKe95]. Everywhere in the superheated liquid, homogenous nucleation of gas bubbles takes now place. The simulation reflects this well at the frame about 15.2ps after irradiation. The void growth further accelerates the removal of matter and larger clusters are ejected from the surface.

A few hundred ps after irradiation the surface region consists of a mixture of liquid droplets and gas phase [LLM03]. Hundreds of nanoseconds after irradiation, the ablated zone relaxes and a crater with modified and often organized surface structures is left behind.

The damage effect and amount of removed matter usually depend on the single-pulse ablation threshold fluence F_{abl} of the material. The processed targets in this thesis are silicon, copper, stainless steel and brass. For silicon $F_{abl} \approx 0.2 J/cm^2$ [JGL+02]. For copper, stainless steel and brass the thresholds are $F_{abl} \approx 0.35 J/cm^2$, $F_{abl} \approx 0.16 J/cm^2$ and $F_{abl} \approx 0.1 J/cm^2$ [NSo10], [MPo00], respectively.

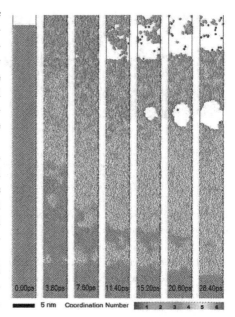

Figure 1.1: *Ablation sequence of silicon. A 100fs, $\lambda=800$ nm, fluence of $0.16 J/cm^2$ laser pulse enters the surface from the top right side and starts at $t = 1$ps with peak intensity at $t = 1.1$ps. Coloring indicates number of nearest neighbors) [HKF+04]. Simulation conditions are close to experiential setup of this work.*

All the described effects take place at laser fluences near the single-pulse ablation threshold. For irradiation by near-infrared laser pulses with sub-threshold fluence, the damage to the surface is significantly less than for fluences equal or above the threshold. In this case the target usually gets irradiated with multiple pulses.

1.2 Laser-Induced Periodic Surface Structures

1.2.1 A Universal Phenomenon

Periodic patterns, generated by laser irradiation at the bottom of ablation areas have been observed since 1965 [Bir65]. These modifications of the solid surface often have the shape of lines (e.g. ripples) and are similar to periodic patterns upon ion beam sputtering or natural occurring structure formations like sand-ripples in deserts and shores, displayed in Fig.(1.2).

Laser-Induced Periodic Surface Structures (LIPSS) (often termed "ripples") have been widely studied on semiconductors, metals and dielectrics. Their appearance on the various materials,

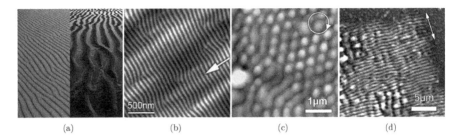

(a)   (b)   (c)   (d)

Figure 1.2: *(a) Sand ripples in desert and seashore from wind & water erosion. p.a.m.f [PJ14], (b) Ripples on silicon produced by ion beam sputtering (arrow indicates direction of ion beam) [ZCF+09], (c) nano-dots produced with circular polarized light on silicon [Var13b] and (d) LIPSS coexistence of ripples and coarser structures on silicon.*

is universal, however, their shape and size depend on the irradiation conditions such as laser polarization and irradiation dose [Var13b], as well as on material properties.

A surface with microscopic irregularity is easier to structure, since the LIPSS start to grow around isolated defects or scratches [SSt82], [RVC08]. It is possible to produce LIPSS with very few laser pulses, however, the influence of scratches on those patterns is large, usually pulse trains of tens to thousands of pulses depending on the laser fluence are necessary to form a pronounced pattern.

Several different types of LIPSS are known, which often coexist in the same irradiated spot. The ripples, are defined as linear periodic patterns with a definite or averaged spatial period length Λ_{LIPSS}. They are further distinguished in three cases. If Λ_{pri} is in the range of hundreds of nanometers > 200nm but below < 1μm, they are in this Thesis, referred to as primary ripples. In the literature they are also known as *LSFL-Low Spatial Frequency LIPSS* ($\Lambda_{\text{LIPSS}} \sim \lambda_{laser}$) [BRK09]. For ripples with period lengths below $\Lambda_{\text{fine}} \leq 200$nm, one speaks of fine-ripples (also known as *HSFL-High Spatial Frequency LIPSS* ($\Lambda_{\text{LIPSS}} \ll \lambda_{laser}$). Period lengths and structures with sizes above $\Lambda_{\text{sec}} > 1\mu$m are here referred to as secondary-ripples or coarser structures, respectively, also known as *grooves*. The size and orientation of the ripples, can be controlled by parameters such as irradiation dose and laser polarization, correspondingly [VCR+07]. Whereas the dose of irradiation I_D is determined as a product of number of pulses and laser fluence.

On semiconductors and metals, the primary ripple pattern is found to grow perpendicular to the direction of laser polarization, while the coarser secondary as well as the fine ripples are often found parallel to the polarization. Upon circularly polarized laser irradiation, LIPSS are characterized by a homogeneously distributed pattern with a lack of linear order. For Silicon, spherical ~100nm sized dots have been reported [VCR+05]. This can be seen on the image in Fig.(1.2(c)). Secondary-patterns and *coarser*-structures can evolve from the ripples when the irradiation dose it high enough.

Λ	Orient.	Period
Λ_{fine}	$\parallel \vec{E}$	($\leq$ 200 nm)
Λ_{pri}	$\perp \vec{E}$	(< 1 μm)
Λ_{sec}	$\parallel \vec{E}$	(> 1 μm)

Table 1.1: *Classification of pattern periods used in this work.*

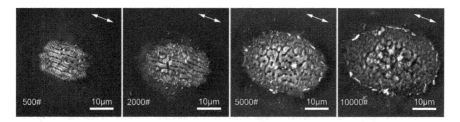

Figure 1.3: *Series of SEM micrographs of LIPSS on Si after different irradiation doses in air. Setup 100 fs pulses with λ = 800nm and a fluence of 2.0J/cm² with. Λ_LIPSS = 650nm later period doubling. Number in lower left corner indicates number of pulses, arrow in upper right corner corresponds to laser polarization.*

In figure (1.3) modified spots on Si(100), generated with different number of applied pulses at equal fluence ($\lambda_{Laser} = 800$nm, $\tau_p = 75$fs), are presented. At the edge of the ablation spot periodic structures with $\Lambda_{LIPSS} = 650$nm, that are orientated perpendicular to the laser polarization are visible. The center of the spot exhibits the secondary pattern, which is parallel to the polarization at a period of $\Lambda_{sec} \approx 2 \, \mu m$. At an increased number of pulses $N > 2000$, ripples in the outer regions as well as the structures in the center become progressively irregular and are finally ablated. The formation mechanism of LIPSS is currently not clear. Two main models, which differ in their physical mechanisms are discussed in literature [BRK09], [GCP+11], [Var13b].

1.2.2 "Static" (Interference) Model

The "static" model describes LIPSS formation as a lithography-like imprint of the structures through an interference pattern, generated between the incident light waves and electromagnetic surface waves [BRK09]. It is supposed that upon irradiation collective longitudinal oscillations

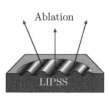

Laser Interference Ablation

Figure 1.4: *Sketches showing the model of the SPP-based process of LIPSS generation: l. excitation of SPP's and coupling of laser energy into them, c. interference of SPP's and EM waves with periods depending on transient index changes, r. spatially varying ablation results in grating-like structures.*

of electrons called *surface plasmon polaritons* (SPP's) can be excited and propagate along the conductor-dielectric interface [Boa82]. During the interaction of laser radiation with the electromagnetic SPP-field, the material's complex refractive index changes locally. This leads to a spatially modulated energy deposition and thus an inhomogeneous ablation process, which forms grating-like structures [SYP+83], [BRK09].

Characteristic features of the theory are for once the dependence of the ripple period length on the angle of incidence θ as well as the laser wavelength λ_{laser} and polarization [GCP+11]:

$$\Lambda_{\text{LIPSS}} = \frac{\lambda_{laser}}{\eta(\tilde{\varepsilon}_m, \varepsilon_d) \pm \sin\theta} . \tag{1.1}$$

The $\pm$ refers to forward and backward traveling surface waves, while the factor $\eta = \lambda_{laser}/\lambda_{SPP}$ is the relation between the laser and surface plasmon wavelength. It describes the real part of the effective index of the surface plasmon mode and further depends on the target optical properties like the permittivity ε_d of a dielectric medium and complex dielectric constant $\tilde{\varepsilon}_m$ for metals [GCP+11]. Typically $\eta \cong 1$ for metals, hence under normal incident angle $\Lambda_{\text{LIPSS}} \sim \lambda$.

The static model is often associated with the early formation stage as well as ripples produced with single pulses. Features such as parallel line formations with period lengths close to the incident laser wavelength can be predicted and well explained. Yet it is not explanatory enough to describe the variety of LIPSS observed.

For example, the splitting and coalescence of ripple lines (Bifurcation and truncation) or cone and "bubble" like structures on dielectrics [Var13b] as well as nano-dots produced by circular polarized light (recall Fig.(1.2(c)) or coarser formations (recall Fig.(1.3)) can so far not be explained by an optical interference effect.

1.2.3 "Dynamic"- (Self-Organization) Model

The consideration of a different physical process was triggered by the observation of many common characteristics of LIPSS and periodic structures formed from instabilities, such as ripples produced on solid surfaces by bombardment with energetic ion beams (i.e., ion beam sputtering IBS Fig.1.2(b)). For instance, similarities were found on the dose dependance (ions energy and irradiation time), the preferential direction (parallel or perpendicular alignment to direction of ion beam), the positive feedback (coalesce of structures at increased ion beam rates) and morphological aspects (bifurcations, dots, cones) [Cos06].

Thus in the "dynamic"-model the LIPSS formation mechanism is based on a self-organization process triggered by surface instabilities. Upon single- or multi-pulse laser irradiation below ablation threshold, the system is driven out of thermodynamic equilibrium resulting in a high-degree of instability.

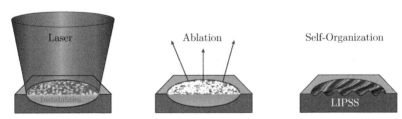

Figure 1.5: *Sketches of ripple formation in the dynamic model. Laser induces high degree of surface instability. The perturbed surface then relaxes through self-organization.*

The material's crystalline order is destabilized and may disappear in the irradiated zone. However, it persists in the surrounding non-affected region [LLS+05], indicating a large gradient of perturbation (recall Fig.1.1). Pump-probe experiments have revealed such transient states of target instability [RCV+07], [Cos06]. As described in section (1.1), a non-equilibrium liquid-like layer forms, which exhibits a corrugated surface morphology.

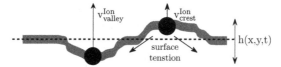

Figure 1.6: *Sketch of a corrugated surface after multi-pulse excitation. Higher ablation velocity of valley ions due to higher next-neighbor density. Surface tension in crests is higher than in valley. Corrugation height is represented by h(x,y,t).*

Due to a higher next neighbor density, the ions in valleys experience a stronger repulsive force, which results in higher escape velocity of the ions. As a consequence, valleys in the corrugated film will be eroded faster than crests. This roughens the surface. On the other hand, thermally activated self-diffusion of atoms simultaneously smoothens the surface [VCR+07]. The competition of both processes drives the instability. A mathematical description of this procedure is adapted from the theory of IBS, where the surface height evolution h(x,y,t), is represented by a rate equation [Sig69], [BHa88]:

$$\frac{\partial h}{\partial t} = -\upsilon \left(\xi, \psi(r), \varepsilon(r) \right) \sqrt{1 + (\nabla h)^2} - K(T)\Delta^2 h , \qquad (1.2)$$

with the erosion (ablation) velocity υ and the self-diffusion coefficient $K(T)$.

The key is the correct calculation of υ, which depends on the material parameter ξ as well as on $\psi(r)$ containing information about spatial laser beam parameters, absorption coefficient, and other energy losses, and $\varepsilon(r)$ describing the energy that results in the erosion at the surface [Var13b]. The dependence on the polarization state of laser light is introduced in υ as well. This yields an equation of motion known as an anisotropic noisy Kuramoto-Sivashinsky [KTs76] equation:

$$\frac{\partial h}{\partial t} = -\upsilon_0 + \nu \Delta h + \lambda \left(\nabla h \right)^2 - K \Delta^2 h + \eta . \qquad (1.3)$$

The first term stands for the initial ablation velocity. The second term describes the ablation rate in linear regime, the third term the non-linear contributions to surface erosion and the fourth term stands for the counterbalancing diffusion due to surface tension. The last term η represents the initial surface roughness. In the linear regime, this means taking the third & fourth term (including K^1) constant, Eq. (1.3) has transient solutions, that visualized show periodic surface structures like ripples.

---

[1]Experimentally, this assumption corresponds to laser fluences well below the ablation threshold, so lattice-perturbations or -temperature do not depend on K

From the linear stability analysis of Eq.(1.3), one can derive the ripple period length $\Lambda_{\text{LIPSS}}(E_a, F)$, as a function of absorbed laser fluence F (i.e. irradiation dose) and activation energy E_a of the atomic self-diffusion:

$$\Lambda_{\text{LIPSS}} = 2\pi \sqrt{\frac{2K}{|\nu|}} = \sqrt{\frac{\frac{2}{aF} \exp\left\{-\frac{E_a}{aF}\right\}}{bF}} \propto \frac{1}{F} \exp\left\{-\frac{E_a}{2aF}\right\} . \tag{1.4}$$

Here, the diffusion rate is assumed to follow an Arrhenius-law:

$$K(T) \propto \frac{1}{k_B T} \cdot \exp\left\{-\frac{E_a}{k_B T}\right\} , \tag{1.5}$$

which also is reflected in the behavior of the period length, displayed in Fig.(1.7). It is important to note, that the surface excitation $k_B T = aF$, described by the temperature, depends linearly on the applied fluence. As well as the erosion coefficient $\nu = bF$ at the surface, which also depends on the fluence. The coefficients a, b are constants of proportionality [Var13b]. Compared to the "static" model one can note an independence of Eq.(1.4) from the laser wavelength λ_{laser}.

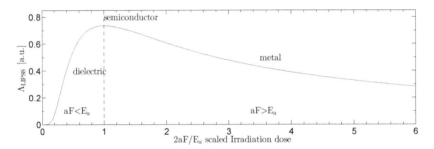

Figure 1.7: *Prediction of general ripple period behavior as a function of applied irradiation dose, in the linear regime of Eq.(1.3). Segments of the curve can be associated with the different materials.*

For dielectrics, Λ_{LIPSS} increases exponentially with increasing irradiation dose[2], the effect is known as a coarsening of the structures [CHR03]. In the case of semiconductors, like silicon, the ripple wavelength does not vary much in the linear regime. If the applied irradiation dose exceeds the binding energy of surface atoms (i.e. activation energy E_a), period lengths are predicted to decrease and become finer. Indeed, this has been found experimentally for metals [RQu10].

In figure (1.8) numerical results for the surface height evolution (Eq.(1.3)) in the linear regime are displayed. The simulation shows the stages from induced instability to the formation of a periodic ripple pattern, as well as a coarsening of periods. The coarsening separates the linear and nonlinear regimes. Outside the linear regime, the surface morphology is determined just by nonlinear terms in Eq.(1.3), that results finally in the collapse of the pattern (not displayed in the simulation).

---

[2]Recall, experimentally the irradiation dose is changed through the number of applied pulses N or laser fluence F: $I_D \propto (F \cdot N)$, or both.

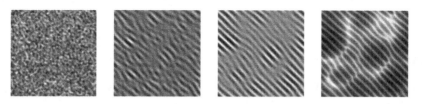

Figure 1.8: *Simulation series of ripple formation (time evolution of Eq.(1.3), upon linear polarized light irradiation. With increasing dose ripple periods increase and coalesce) [Var13a]*

A similar behavior is observed experimentally. With increasing irradiation dose a positive feedback develops and the ripples start to coalesce, exhibiting more complex, irregular, coarse features. The progressive development of the periodic surface structures is only observed in an appropriate range of the applied irradiation dose.

Comparison

A direct comparison of both LIPSS formation models show their main differences.

The *static*-model needs a monochromatic and spatially coherent light source (fundamental properties of laser), in order for an interference effect to take place on the surface. Furthermore, if the relation between laser and SPP wavelength is about $\eta \cong 1$(metals), structures with periods close to the laser wavelength will form under normal incident angle.

Static-model:

- depends on monochromatic light
- depends on spatial coherence of light
- $\hookrightarrow$ interference effect on surface
- $\boxed{\Rightarrow \Lambda_{\text{LIPSS}} = f(\lambda_{laser}, \lambda_{laser}/\lambda_{SPP}, \theta)}$

Dynamic-model:

- independent of light wavelength $\Lambda_{\text{LIPSS}} \neq f(\lambda)$
- independent of spatial coherence
- depends on irradiation dose
- $\hookrightarrow$ induced surface instability
- $\boxed{\Rightarrow \Lambda_{\text{LIPSS}} = f(I_D(F \cdot N))}$

The *Dynamic*-model, based on induced surface instability, has no requirements on the chromatic or coherence properties of the light source. Period lengths are only influenced through the applied irradiation dose.

Both models require an ultra-short pulse duration of the used light pulses.

Chapter 2

Generation of White Light Supercontinuum

Supercontinuum generation (SCG) is a process, where laser pulses with narrow spectral bandwidth (quasi-monochromatic) are converted to pulses with very broad spectral bandwidth.

If a high-peak-intensity pulse is focused into a transparent nonlinear material, the output pulse will, on a screen, appear as a white light flash, regardless of whether the input pulse is in the near IR or near UV range. The spectral broadening is mainly driven by the nonlinear optical effects of self-phase modulation (SPM), which leads to the generation of the new frequencies, and self-focusing (SF), which is responsible for very high peak intensities.

This phenomenon was first discovered with ps pulses by R. Alfano & S. Shapiro [ASh70] and remains an active field of research with applications for example in time-resolved spectroscopy [Ekv00], coherence tomography [HLC+01], optical parametric amplification (OPA) [WYa97].

Continuum Generation in Sapphire

In this work the material used for SCG is an Al_2O_3 crystal often referred to as corundum or sapphire. It is known to afford a broadband SC and possesses a high damage threshold. Al_2O_3 is a centrosymmetric medium (i.e. it possess a center of inversion) that furthermore has no intrinsic free carriers, is non-conductive and non-magnetizable. These properties are essential to the SF and SPM effects involved in the CG.

2.1 Nonlinear Optical Effects

The interaction between an electromagnetic (EM) wave and the crystal, strongly depends on the strength of the electric field $\vec{E}(\vec{r}, t)$.

If the amplitude $|\vec{E}(\vec{r}, t)|$ of the electric field is much less than the inner Coulomb field of an atom (typically in the order of 10^9 V/m), the displacement of the electrons is small and the restoring force can follow electric-field linearly (harmonic oscillator). Thereby, the induced electric dipole moments are proportional to $\vec{E}(\vec{r}, t)$, hence, the response is linear and results in a polarization $\vec{P}(\vec{r}, t)$ of the medium, which is described by:

$$\vec{P}(\vec{r}, t) = \varepsilon_0 \, \underline{\chi}^{(1)} \, \vec{E}(\vec{r}, t) \ . \tag{2.1}$$

The proportionality factor $\underline{\chi}^{(1)}(\omega)$ is a 2nd-rank tensor and known as the linear susceptibility of the medium. It depends on the frequency of the incident EM field. In this chapter an instantaneous response of the medium without a reaction time, is assumed [Boy03]. In an isotropic material $\chi^{(1)}$ is a scalar quantity and related to the refractive index n of the material:

$$n = \sqrt{1 + \chi^{(1)}} \ . \tag{2.2}$$

If $|\vec{E}(\vec{r}, t)|$ gets closer to the Coulomb field, the induced electric dipoles in the medium cannot respond linear anymore resulting from the anharmonic potential between the electrons and the nucleus, thus, higher order terms contribute to the polarization:

$$\vec{P}(\vec{r}, t) = \vec{P}^{(1)} + \vec{P}^{(2)} + \vec{P}^{(3)} + \ldots = \varepsilon_0 \left[\underline{\chi}^{(1)} \ \vec{E} + \underline{\chi}^{(2)} \ \vec{E}\vec{E} + \underline{\chi}^{(3)} \ \vec{E}\vec{E}\vec{E} + \ldots \right] \ . \tag{2.3}$$

For sapphire with a centrosymmetric crystal structure, the even order susceptibility terms in Eq.(2.3) vanish [Boy03], so that only odd-order terms contribute to the polarization and cause distinct nonlinear effects.

2.1.1 Self-Focusing

The third-order nonlinear polarization is responsible for various phenomena. It can be related to processes like four-photon interactions (third harmonic generation, THG, and four wave-mixing), or nonlinear response effects of the material. One of the most important effects is the nonlinear contribution to the refractive index, which gives rise to various self-action effects of the laser beam like self-focusing in a crystal.

Higher-orders of nonlinearity can be neglected, since their strength of susceptibility decreases $\chi^{(n)}/\chi^{(n+1)} \approx 10^{12}$ [Boy03]. For demonstration it's easier to further assume a linearly polarized field in a dispersionless, isotropic medium. Then, the tensor character of $\underline{\chi}^{(3)}$ and the vector notation can be suppressed, leading to the medium total third-order response of:

$$P = P^{(1)} + P^{(3)} = \varepsilon_0 \left[\chi^{(1)} + \frac{3}{4} \chi^{(3)} |E(\omega)|^2 \right] E(\omega) \ . \tag{2.4}$$

The factor of $3/4$ rises from the summation over all permutations of the four photon interaction, as well as the assumed simplifications [Boy03]. The expression in the brackets is also referred to as the effective susceptibility $\chi_{eff} = \chi_L + \chi_{NL}$, containing a linear and nonlinear term. Hence, a nonlinear contribution is added to the material's refractive index in Eq.(2.2):

$$n^2 = 1 + \chi_L + \chi_{NL} = n_0^2 + \frac{3}{4} \chi^{(3)} |E(\omega)|^2 \ . \tag{2.5}$$

Equation (2.5) can be rewritten and it can be shown that the nonlinear contribution is significantly less than the linear one and the following estimation can be applied.

$$n = n_0 \cdot \sqrt{1 + \frac{\chi_{NL}}{n_0^2}} \overset{\chi_{NL} \ll n_0^2}{\approx} n_0 \cdot \left(1 + \frac{1}{2} \frac{\chi_{NL}}{n_0^2} \right) \tag{2.6}$$

With the intensity $I = \frac{1}{2} \varepsilon_0 c n_0 |E|^2$ the nonlinear susceptibility becomes $\chi_{NL} = \frac{3}{2} \frac{\chi^{(3)} I}{\varepsilon_0 c n_0}$. Hence, the total refractive index is finally expressed as:

$$n \approx n_0 + \frac{3}{4} \frac{\chi^{(3)}}{\varepsilon_0 c n_0^2} I =: n_0 + n_2 I \ . \tag{2.7}$$

The quantity $n_2 = \frac{3}{4} \frac{\chi^{(3)}}{\varepsilon_0 c n_0^2}$ is the induced nonlinear refractive index. In Al_2O_3, the refractive index has a value of n_0=1.76 (@800nm) and its nonlinear contribution n_2=2.9$\times 10^{-16}$cm^2/W [Boy03]. Thus, a laser with an intensity of 1TW/cm^{-2} can produce a refractive index change of 2.9$\times 10^{-4}$ in Al_2O_3. Now, this may seem not much, but can have a significant effect.

For a Gaussian laser beam, the spatial intensity profile results in a spatial variation of refraction index, which is higher in the center than at the wings. This variation in refraction index acts as a positive lens (self-focusing) and the beam converges and eventually collapses in one point.

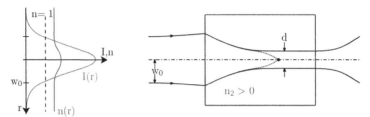

Figure 2.1: *r. Schematics of the intensity profile of a real laser beam and the corresponding refraction index. l. Beam diameter and self-focusing effect in a nonlinear optical medium.*

The occurrence of self-focusing, surprisingly, depends on the beam's power, rather than its intensity, and takes place only if a critical power is reached. In the theory of SF, J.H. Marburger [Mar75] calculated a threshold power for a Gaussian continuous wave (CW) laser beam with a wavelength λ_{laser}:

$$P_{crit} = \frac{3.77\lambda_{laser}^2}{8\pi n_0 n_2} \ . \tag{2.8}$$

For a CW-laser with λ_{laser}=800nm in Al_2O_3, this would equal a critical power of $P_{crit} \approx$ 1.88MW. However, for an ultra-short pulse duration, it has been reported that this threshold is considerably increased due to a temporal spread of the pulse (group velocity dispersion (GVD)) in a dispersive medium such as Al_2O_3 [CPe92], [SCo94], [LWM+94].

If the applied power equals the critical power, another self-action effect, the so called *self-trapping* of light, occurs. Here, a defocusing balances the SF, preventing the beam from collapsing in one point and thus damaging the material. In this case, the beam acts as its own waveguide, propagating with a finite diameter d through the medium. It is reported that the beam in sapphire shrinks to a diameter of 20-25μm [BCh99], [YLW00], [JSM+05].

The main cause of self-trapping and, thus, a limitation of self-focusing, is the generation of free electrons inside the medium [BCh99]. Thereby, a negative change in the index of refraction is induced, through the free-electron density N_e:

$$\Delta n_e = -\frac{2\pi e^2 N_e}{n_0 m_e \omega_{laser}^2}, \tag{2.9}$$

where ω_{laser} is the laser frequency. The main mechanism of free-electron generation in Al_2O_3, with femtosecond laser pulses, is multi-photon excitation (MPE).

Only in the central part of a Gaussian pulse (both spatial and temporal), where the intensity exceeds 10^{12} $[Wcm^{-2}]$, free-electrons can be generated in sapphire, through four-photon absorption. If the free-electron density reaches 10^{17}-10^{18} $[cm^{-2}]$ [JSM+05], [BCh99], Δn_e cancels out the nonlinear refraction index $n_2 I$ and self-focusing and defocusing come to an equilibrium. The lower intensity wings of the pulse are defocused due to diffraction.

For a pulse, which greatly exceeds the critical power P_{crit} for SF, the trapped beam, in transverse direction, breaks up in many separate beams (filaments), each of which contains the power P_{crit} and undergoes their own self-trapping effect. This is called *laser beam filamentation*. A Gaussian input beam becomes randomly modulated in intensity, overall reflecting the initial envelope with spots of higher intensity in the center. These spots, also referred to as *hotspots*, fluctuate over time in intensity, but in average keep their transverse position.

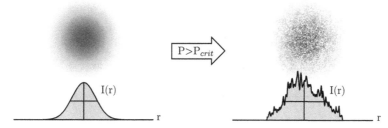

Figure 2.2: *Schematics of a Gaussian intensity distribution before (left) and after laser beam filamentation (right) as well as the corresponding transverse intensity profile. The beam breaks up in a large number of individual components each of which carries the power P_{crit}. p.a.m.f [Wik14].*

The nonlinear self-action effects of self-focusing, self-trapping and multi-beam filamentation are essential for the generation and enhancement of white-light supercontinuum. It is widely accepted that self-focusing triggers self-phase modulation.

2.1.2 Self-Phase Modulation

Self-phase modulation (SPM) is the dominant process for the generation of new frequencies from a fundamental laser frequency ω_0 in a nonlinear optical (NLO) medium such as sapphire. The mathematical description for the simple theory of SPM [YSh84], [Alf06] is straight forward and can be demonstrated as follows. The electric field of an incident laser pulse propagating in z direction can be expressed as,

$$E_{in}(z_0, t) = E_0(z,t)\, e^{i(knz-\omega_0 t)} + c.c. = E_0(z,t)\, e^{i\phi_0} + c.c. \quad (2.10)$$

where $k = \lambda/2\pi$ is the wave-number and n the material's refractive index. For simplicity, only the forward traveling wave is considered. Upon entering the NLO-medium of length L, the phase experiences an intensity dependent shift, in its propagation from 0 to L.

$$E_{out}(L, t) = E_0(L,t)\, e^{i(kn_0 L + kn_2 I(t)L - \omega_0 t)} =: E_0(L,t)\, e^{i(\phi_{NL}(L,t)+\phi_0)} \quad (2.11)$$

Note the time-dependence of the additional phase $\phi_{NL}(L, t)$, coming from the time-dependent intensity $I(t)$. The induced phase change $\Delta\phi = \phi_{NL} - \phi_0$, therefore, arises only from the intensity dependent refractive index of the material (Eq. 2.7). Since the frequency of the pulse's electric field is given by $\omega = -[\partial(\phi_{NL} + \phi_0)/\partial t]$, the phase modulation leads to a frequency modulation, with $k = \omega_0/c$,

$$\omega = \omega_0 - n_2 \frac{\omega_0}{c} L \frac{\partial I(t)}{\partial t} . \qquad (2.12)$$

This situation is displayed in Fig. (2.3) for a τ_p=100fs pulse of a Gaussian intensity profile, with a peak intensity of 100TW/cm^2 in the temporal domain, that undergoes a frequency shift in a 5mm long Al$_2$O$_3$ crystal. The leading edge (the left-hand wing) is thereby shifted to lower

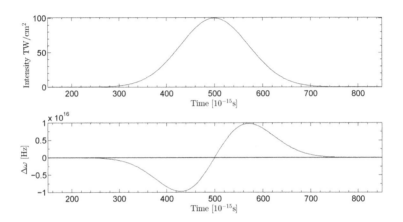

Figure 2.3: *(upper) temporal intensity profile a τ_p=100fs Gaussian laser pulse, (lower) frequency shift of the fundamental frequency ω_0 in a NLO-medium*

frequencies (i.e. longer wavelengths), whereas the trailing edge experiences a shift to higher frequencies (i.e. shorter wavelengths). The consequence is a broadening of the pulse spectrum $S(\omega)$ resulting from the time-dependent intensity slope, which can be calculated from the Fourier transformation. The corresponding spectrum of the output pulse is displayed in Fig.(2.4):

$$S(\omega) = |E_{out}(t)|^2 = \left| \int_{-\infty}^{\infty} E_0(t) \, e^{-i\omega_0 t - i\phi_{NL}(z,t)} \, e^{i\omega t} dt \right|^2 . \qquad (2.13)$$

It shows a frequency broadening of about 2×10^{16}Hz in comparison to the fundamental laser frequency. Thus, the spectrum of the pulse spreads now from about 600nm up to 1000nm. The quasi-periodic oscillations of the spectrum, arise from symmetric phase change $\Delta\phi$, which further results from the symmetric intensity profile. Two newly generated waves (ω_1 & ω_2) with equal frequencies but different phases, can interfere constructively for $\Delta\phi_{12} = 2\pi$ or destructively for $\Delta\phi_{12} = \pi$. This example serves as a qualitative picture for demonstration of the SPM process.

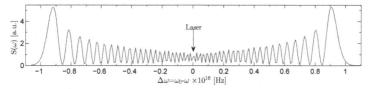

Figure 2.4: *Theoretical power spectrum of the phase modulated pulse with Gaussian temporal intensity distribution, which assumes an instantaneous response of the material.*

The red-shifted part of the spectrum is often referred to as Stokes broadening, while the blue-shifted part is referred to as anti-Stokes broadening. The displayed spectrum has a Stokes-anti-Stokes symmetry. Another consequence of the SPM is that the frequency of the output pulse changes over time, which means the pulse exhibits a high linear chirp.

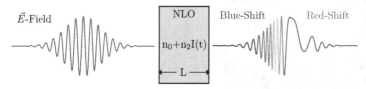

Figure 2.5: *SPM of a pulse (left) in a NLO-medium, results in a linear chirped output pulse (right).*

In this example of SPM an instantaneous response of the medium and, therefore, of n_2 was assumed. So, the phase modulation $\Delta\phi$ is directly proportional to the intensity variation $I(t)$. Generally, this is not the case. The medium's finite response time causes a distortion of the phase modulation $\Delta\phi$, which can result in a Stokes-anti-Stokes asymmetry, even if $I(t)$ is symmetric. Most spectra observed exhibit a strong anti-Stokes broadening (blue-shifted frequencies). This can qualitatively be explained as follows. The intensity dependence of the phase velocity $-\partial\phi[I(t)]/\partial t$ is responsible for the SPM, however, the intensity dependence of the group velocity (Eq.(2.14)) leads to the *self-steepening* of the pulse.

$$v_{\text{group}} = \frac{c}{n(I) + \omega\frac{\partial n(I)}{\partial\omega}} \tag{2.14}$$

From Equation (2.14) it follows, that the peak intensity of the pulse will have a smaller group velocity than its wings, because of the inverse proportionality to $I(t)$. Thus, the peak of the pulse falls away from the leading edge into the trailing edge, resulting in a steep intensity edge towards the back of the pulse. This higher slope in intensity will result in more blue-shifted frequencies towards the back of the pulse, than red-shifted and thus, in an asymmetric spectrum.

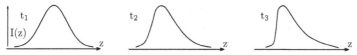

Figure 2.6: *Self-steepening of an incident Gaussian pulse with $t_1 < t_2 < t_3$. Typically the trailing edge exhibits a steepening, upon propagation through a NLO-Medium.*

Self-steepening, therefore changes the shape of the pulse. Self-phase modulation can also be considered in the spatial domain. In that case, the transverse intensity variations of the pulse lead to a nonlinear contribution to the refractive index $n_2 I(r)$ in transverse direction. Thus, the induced phase change $\Delta\phi(r, z)$ will cause a distortion of the wavefront and upon self-focusing, a drastic modification in the beam cross-section. As a consequence, different frequency components of the beam diffract into cones under different angles. If projected onto a screen, the continuum appears as a round white disk often surrounded by a distinct concentric rainbow-like pattern, also called *conical emission*, which exhibits a large divergence angle.

 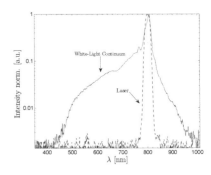

Figure 2.7: *White-light continuum from pump-laser ($\tau_p = 75fs$, $\lambda = 800nm$) with a Gaussian intensity profile in Al_2O_3, projected onto a screen. Visible is white-light central disk and conical emission*

Diffraction is one explanation of the conical emission, but it has also been suggested that the effect could arise from four wave-mixing [XYA93], since the high divergence would satisfy the phase-matching condition. In Fig. (2.7) the output of a SC generated with femtosecond pulses (τ_p=75fs & λ=800nm), with a peak power of P_{peak}=1.56 GW per pulse, in Al_2O_3 is shown. The antisymmetric spectrum, taken in the center of the white-light disk, spreads from 450nm to 1000nm. It clearly shows a strong anti-Stokes broadening with more blue-shifted frequencies. The peak intensity of the WLC has about a tenth the intensity of the pump-laser.

The process of continuum generation only starts if a certain threshold is reached. Experimentally, it has been found that this threshold power equals the critical power for SF [TIm09], [BCh99]. However, during the SPM the new frequency components suffer a temporal lack due to group velocity dispersion in the medium. The pulse inevitably becomes longer and thus its peak intensity reduces, so that a higher input power is needed for SF to start. How much the pulse temporally spreads depends on the frequency components. In [QWW+01] laser pulses (τ=130fs & λ=800nm) generated a WLC in H_2O, which spread from about 400nm to 1000nm. The individual frequency components showed an average WL pulse duration of 137fs. The WL pulse thereby propagated 1cm though the H_2O.

This shows that the WLC pulse stays an ultrafast light pulse during its generation and subsequent propagation in the NLO-medium. It thereby remains attractive for many physical applications that require ultrafast laser pulses such as femtosecond laser ablation.

Chapter 3

Instrumentation and Experimental Setup

3.1 Laser System and Beam Quality

The setup for the experimental investigations is based on a commercial "solid state" femtosecond laser system from Spectra Physics, USA. The system provides femtosecond laser pulses from a mode-locked Titan:sapphire (laser medium a sapphire crystal doped with Titan ions Ti^{3+}:Al_2O_3) oscillator (*Spectra Physics Tsunami, Model 3960*) with a central wavelength of λ_{laser}=800nm and a spectral width at the full width half maximum (FWHM) of $\Delta\lambda$=15nm.

The pulse duration τ_p and shape in the temporal domain were determined with a multi-shot autocorrelator (*Spectra Physics, Model 409*). A detailed explanation of these quantities is given in the Appendix (A). The Tsunami oscillator provided laser pulses of a Gaussian intensity distribution in the temporal domain with a pulse duration at FWHM of $\tau_p \approx$75fs. The energy per pulse lies in the range of nJ, which is not enough for continuum generation or ablation of solid surfaces. Thus, the pulses were amplified using the chirped pulse amplification (CPA) technique [SMo85] in the *Spectra Physics Spitfire* regenerative amplifier. The Spitfire then released pulses with a maximum pulse energy of $E_p \approx 350\mu J$, at a repetition rate of 1000 pulses per second (f_{rep}=1kHz). This train of pulses, hereafter referred to as pump-laser beam (or beam), is horizontally polarized, with respect to the worktable. The Spitfire amplifier releases a TEM_{00} transverse electromagnetic mode, also known as Gaussian mode. So the femtosecond pulses leaving the amplifier should exhibit a Gaussian intensity distribution in the spatial domain and thus a Gaussian beam propagation. These quantities as well as the quality of the beam (how close to the ideal Gaussian case) were determined with a monochrome (*EHD, Model 12V5HC*) CCD-camera. A detailed explanation of the determination is presented in Appendix (B) as well.

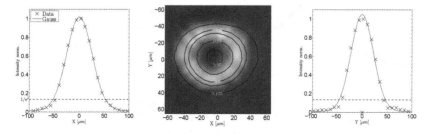

Figure 3.1: *Transverse intensity profile and intensity cross-sections of the laser beam in the focus of a f=+300mm lens. Isohypses in central plot show:$1/e^2$, $1/e$, 0.5, 0.75 and 0.95 of I_{max}.*

As shown in Fig.(3.1), the focused beam exhibits a Gaussian intensity profile in transverse direction as expected. The transverse distribution, in the focus of a f=+300mm lens, is slightly elliptical with a major axis in the x-direction (horizontal) and corresponding waist sizes in the focus of $\bar{w}_{0x,300}$=45.11μm and $\bar{w}_{0y,300}$=40.99μm in y-direction (vertical). The quality of the beam with an M^2 factor of $M_x^2 = 1.103$ and $M_y^2 = 1.225$ is close to the ideal case of Gaussian beam propagation (M^2=1.0). Thus, the beam has a quality of 91±4% and 82±8% for the corresponding x and y-directions. The laser system operated under very stable conditions of humidity and temperature, controlled by an air conditioner which kept the room temperature at $23°C$. This ensured, a stable pulse energy ($\Delta E_0/E_0 \approx 3\%$) and high long-term pointing stability.

3.2 Experimental Setup

A complete picture of the experimental setup for the investigation of the white-light supercontinuum and subsequent ablation with ultrafast white-light, with all components in the beam path is presented in Fig.(3.2).

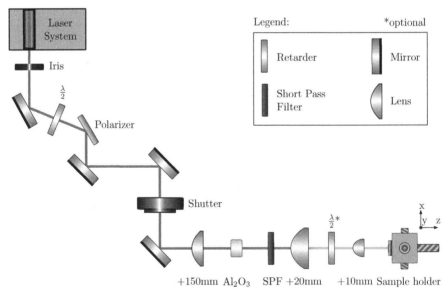

Figure 3.2: *Sketch (not to scale!) of the complete experimental setup with all components in the beam path. The general and individual reoccurring components are listed in the upper-right legend.*

Directly after the output from the laser system the beam is spatially filtered, blocking eventual diffraction orders and selecting only the central part of the Airy pattern was selected with an iris. This produces a beam of w_x=2.02 mm width in the horizontal- and w_y=3.56 mm the vertical-direction (regarding $1/e^2$). Both quantities also determined with a monochrome CCD-camera.

The intensity of the beam could be attenuated with a half wave-plate retarder, followed by a thin-film linear polarizer. The p-polarized beam (with respect to the polarizer), is sent under the Brewster angle to the polarizer. The polarizer, then passes only portions of its specified vertical (workspace) direction. By changing the incident polarization plane with the $\lambda/2$ wave-plate to vertical (workspace), light with the highest intensity leaves the polarizer. In between, the polarization plane can be changed continually and a desired intensity can be selected.

After the intensity attenuation arrangement, the beam is guided by three more silver coated mirrors to the setup for white-light characterization and LIPSS generation. The beam enters the setup through a mechanical shutter, which is activated by a trigger box synchronized with the laser chain. It can be programmed to open and transmit a desired number of laser pulses.

After the shutter, the laser beam is focused by a $f=+150$mm plano-convex lens into the Al_2O_3 crystal. The crystal has a cylindrical shape with a length of $l=5.05$mm and a diameter of $d=3.05$mm. The top and bottom of the crystal are highly polished, while the lateral area is roughened.

A short pass filter transmits the anti-stokes part of the generated white-light continuum and blocks the pump-laser beam. Behind the filter, a telescope arrangement is placed to tightly focus the diverging white-light onto the sample. It consists of one $f=+20$mm plano-convex lens with a large diameter of $d=30$mm and a $f=+10$mm plano-convex lens with a small diameter of $d=6$mm. In between the lenses of the telescope a $\lambda/2$ wave-plate can be mounted, to change the orientation of the polarization plane. The last part of the setup is the custom made sample holder mounted on a precision translation stage. With this the sample can be translated in the x,y,z-direction with a precision of 100μm.

Power-detector and Fluence determination

The power of the laser beam and subsequent of the white-light continuum, is measured with a *Gentec PS-310* power-meter. It is basically a thermopile, which absorbs the incident light and converts it to heat. This results in a temperature difference between the hot absorber and the cold heat sink.

The temperature difference causes the thermopile to generate a voltage, which then can be measured. The voltage is proportional to the beam power and laser fluence $V \propto P \propto F$. The average power of the beam P_{ave} is simply the measured voltage divided by a calibrated reference Voltage, $P_{ave} = V_{\text{meas.}}/117mV$ [W]. With the average power P_{ave} and the irradiated area A (effective area of the beam profile), the peak laser fluence in $[J/cm^2]$ can be calculated as follows,

$$F = \frac{P_{ave}}{f_{rep} \cdot A} \tag{3.1}$$

To obtain the peak intensity in $[W/cm^2]$, equation (3.1) is simply divided by the pulse duration τ_0 (at the FWHM). The energy per pulse in $[J]$ is given by $E_0 = P_{ave}/f_{rep}$. The thermodynamic nature of the detector often causes a large random error, especially in lower power ranges. Thus, during a measurement the detector needs to be irradiated at least 45s to go into saturation.

3.3 Short-Pass Filters

In the experiments carried out in this work, it is very important to reject the λ_{laser}=800nm of the pump-laser beam from the white-light continuum. This will be essential, especially for the laser induced periodic surface structures, since the influence of a continuum without coherent contribution in the LIPSS formation process will be investigated.

For this task, two different short-pass filters and one dielectric turning mirror at 800nm were used. The mirror will be further referred to 800nm-filter.

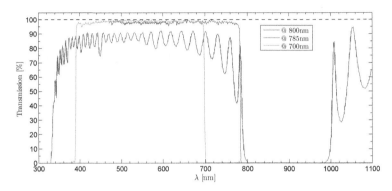

Figure 3.3: *Short-pass filters transmission curves (normal incidence), with cut-offs at (@) 800nm, 785nm, 700nm, used to block the fundamental pump-laser line. (800nm dielectric mirror), [CVI14], [Com14], [Inc14].*

In figure (3.3) the transmission curves of all 3 filters are displayed. The first one (magenta colored graph), is the dielectric mirror from *CVI Laser Corporation*. It is specified to have a reflection of 99.9% in the range of (800nm-990nm) to turn an ultrafast laser beam. The transmission graph shows the familiar modulation of such mirrors, arising from the dielectric coated layers with different refractive indices, in order to reach the high reflection for a defined wavelength. As a filter, it transmits from 330nm up to 790nm with an average of T_{ave}=80.4% and completely cuts-off at 800nm and further up transmits from 1000nm again. The maximum transmission lies at around T_{max}=92.5%. The displayed graph is the specified transmission under an incident angle of θ_{inc}=0°, with respect to the mirror plane.

The second filter, further named as 785-filter (red colored graph), is a Raman-short-pass filter from *Laser Components*, with an average transmission of T_{ave}=95% in the range of 470-779nm and cut-off at 785nm. Laser Components only provided data down to 470nm. It exhibits fluctuations in the transmission of about ±5%, but does not show a major modulation like the dielectric mirror. For the blocked out wavelength range the transmission is lower than $T < 10^{-6}$%

The third one, referred to as 700nm-filter is a premium short-pass filter from *Thorlabs*, with ultra-steep cut-off at 700nm and an average transmission of T_{ave}=98% in the range of 390-696nm. In the blocked wavelength range the transmission is below $T < 10^{-5}$%.

3.4 Sample- and Spectral-Analysis Devices

In this section, the instrumental devices that are used in this work to analyze the processed samples and to measure the spectrum of the laser-beam and the white-light continuum, will be explained in a short way.

Stellar-Net EPP2000 UVN-SR

The spectrometer that was used for the various spectral measurements is an *EPP2000 UVN-SR* portable fiber optic spectrometer from *StellarNet*. The incoming light is captured by a 25μm slit, coupled with a 1000μm diameter single core fiber input cable. Subsequently the light is dispersed by its wavelength in a Czerny-Turner grating spectrograph [CTu30] and recorded by a 2048 pixel CCD-chip, which is most sensitive between 400nm and 800nm [Tos14]. The specification UVN-SR stands for super-range since its spectral range spreads from 200nm-1100nm, and thus covers the ultraviolet (UV), visible (VIS) and near-infrared (NIR) parts of the electromagnetic spectrum. The spectral data were recorded with a personal computer via USB interface.

SEM Zeiss EVO 40

To generally analyze the laser induced surface modifications, as well as to determine the lateral structure sizes, a Zeiss EVO 40 scanning electron microscope (SEM) was used. The SEM produces an image of a sample by scanning it with a focused beam of primary (exciting) electrons. The incident electrons interact with surface layer electrons in the sample and eject them from the surface. Such secondary electrons, created at a topographic peak have a greater chance of escaping than secondary electrons created in a topographic hole. The SEM images are presented in gray-scale. Brighter areas represent topographic peaks, darker areas topographic holes. The setup with the sample is placed in a high vacuum chamber. Lateral resolution down to a few μm can be reached. An advantage of the Zeiss EVO 40, is that the byproduct X-ray spectrum (produced during the electron bombardment) can be used for *Energy Dispersive X-ray Spectroscopy* (EDX), to obtain information about the chemical composition of the surface. This is presented in section (5.5). However, SEM-images are not able to give quantitative information about the surface height profile.

AFM Solver P47H

In order to get information about the surface height profile of the LIPSS, a *Solver P47H*, atomic force microscope was used (AFM). It consists of a tip (nanometer apex size), that slides over the surface, whereas the interaction forces between the tip and atoms of the sample surface cause a deflection of the lever on which the tip is mounted. A change of topography leads to a change in the deflection of the lever. Thereby a three-dimensional image can be obtained by scanning the sample surface. Surface heights of up to 2μm in an area from 1$\mu m \times$ 1μm to 50$\mu m \times$ 50μm can be measured. The great advantage of the AFM is that the sample analysis can be performed in air. All AFM images in this work will be shown in a gold-scale.

Chapter 4

Characterization of White Light Supercontinuum

In this chapter, the measurements of the white-light continuum characteristics in the spectral, spatial and temporal domain are presented and discussed. Of major scientific interest, is for once, the behavior of the white-light continuum to the applied input power of the pump-laser beam. The goal is to later obtain an efficient and possibly flat spectrum (here, moderate slope in narrow spectral intervals), with a large spectral broadening and sufficient white-light fluence F_{WL}, for the formation of LIPSS on the samples. The investigation also includes the determination of the output power of the filtered spectra after the three short pass filters.

Secondly, how does the transverse intensity profile look like and what spectral differences does it may show for larger input powers? This information is always essential for laser ablation in order to compare with the size and shape of the ablation craters and determine the white-light fluence.

And, lastly, does the generated white-light continuum preserve the pump-lasers polarization and high degree of coherence, or does it exhibit a reduced coherence in the spatial and temporal domain, or none at all?

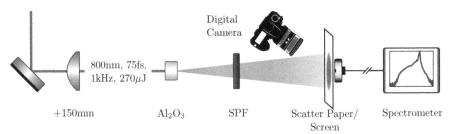

Figure 4.1: *Sketch (not to scale!) of the general setup for the investigation of spectral, spatial and temporal properties of the white-light continuum. (SPF) short-pass filter.*

The investigation of the white-light beam characteristics is carried out with a simplified version (Fig. 4.1) of the complete experiential setup (Fig. 3.2) .

In general the pump-laser provided 75fs pulses at 1kHz repetition rate, with a central wavelength of $\lambda_{laser} = 800nm$. At the position of the crystal the pump laser pulse energy was roughly $E_0 \approx 270\mu J$. The actual energy per pulse was always measured right behind the +150mm-lens and is for each individual experiment given in the corresponding section.

For the spectral measurements of the generated continuum, a regular piece of paper or sometimes lens cleaning paper, was used to scatter and dim the white-light, in order to safely record the spectrum and not to damage the spectrometer.

During the investigation of the spatial properties, the intensity profile was captured with the monochrome CCD camera. For real color images, the white-light continuum was projected onto a screen and images of the disk shaped white-light spot were taken with a Canon EOS 650D digital single-lens reflex (DSLR) camera.

4.1 Onset of Continuum Generation

The continuum is generated by focusing the pump-laser with an +150mm-lens into the Al_2O_3 crystal. Thereby, the crystal is placed directly into the focused area of the pump-beam, so that the beam waist would theoretically be located in the center of the crystal. With the crystal at this position, it was found that the resulting continuum is brightest (highest intensity) and the highest white-light pulse energy, for a certain applied input power, could be obtained.

In the next step the the pump pulse energy, at which the continuum generation starts, is measured. The minimum energy per pulse, that can be attenuated in the experiments is about $E_0 = 4\mu J$, at the position of the crystal. Hence, for the Gaussian intensity profile, the peak power is P_{peak}=53MW. The critical power for self-focusing in Al_2O_3 is $P_{crit} \approx 1.88$MW, for CW-laser with Gaussian intensity profile (Eq.2.8). This value is supposed to be considerably higher for ultra-short pulses due to GVD (recall, chapter 2.1.1). The peak power of P_{peak}=53MW, seems to already exceed P_{crit} by far, however, no white spot could be seen on the screen.

The pump-energy of the beam was now slowly increased until at a threshold of E_{wl}=17.7μJ, a dim colored disk started to appear on the screen (Fig.4.3(a)), which could be seen by the naked eye. This threshold corresponds to a peak power of $P_{peak,wl} = 230.7$MW, indicating the critical power for self-focusing with ultrashort pulses. [BCh99].

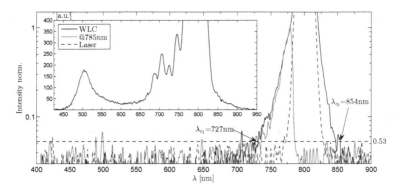

Figure 4.2: *White-light spectrum at the onset of continuum generation E_{wl}=17.7μJ, recoded at low integration time setting of the spectrometer. (Inset) Spectrum recorded without filter and higher integration time (higher number of processed pulses).*

In figure (4.2) the broadened spectrum of the pump-laser pulse at the onset of continuum generation is displayed. The spectrum shows that the pump laser is indeed spectrally broadened and spreads from 727nm $\leq \lambda_{wl} \leq$ 854nm. Beyond these limits the spectrum cuts-off and only noise is detected[1]. However, the actual broadened range is definitely larger, since the spot (Fig.4.3(a)) also exhibits short wavelength colors ($\lambda <$ 700nm).

For a higher integration time setting (higher number of processed pulses) of the spectrometer, wavelengths of low intensity could be recorded. The revealed spectrum (inset Fig.4.2) reaches down to about 450nm with a peak at 500nm. The peaks close the the pump-laser wavelength showed large fluctuations in intensity and position during the measurement, since the spectrometer was almost completely exposed to the light, and thereby recorded also the intensity fluctuations (no scatter paper). Nevertheless, an overall tendency to an asymmetric broadened spectrum with stronger anti-Stokes side can already be found.

(a) $E_0 \approx 17.7\mu J$ ($1.00E_{wl}$)   (b) $E_0 \approx 22\mu J$ ($1.25E_{wl}$)   (c) $E_0 \approx 51.1\mu J$ ($2.88E_{wl}$)

Figure 4.3: *Part one of the white-light spot evolution projected onto a screen at different input pulse energies E_0 (E_{wl} measured energy threshold of continuum generation). No filter was used to block pump-laser.*

Spatially the spectrum seems to be spreading from longer wavelengths ($\sim$ 650-850nm) on the left hand side to the shorter wavelengths ($\sim$450-500nm) on the right side (visible in Fig.4.3(a)). An explanation of this effect is given in the spatial properties section 4.4. The residual pump-laser spot is visible as a faint red dot (indicated by arrow). The white-light spot exhibits a much larger divergence angle of $\theta_{wl} = 5.27° \pm 1.1°$, than the pump-laser spot $\theta_{laser} = 1.1° \pm 0.26°$), in the farfield of the +150mm-lens (Fig.4.3(a)). Here, and in the following investigation, the divergence angles were obtained by taking the ratio of the half disk diameter to the distance from the center of the crystal to the screen. The diameter was measured with an precision caliper by dark-adapted eyes. This technique has obvious limitations, but still provides valuable information. To compensate the lack of precision an uncertainty error of $\Delta d \pm 4$mm was given to the disk diameter.

At a pulse energy of $E_0 \approx 22\mu J$ ($1.25E_{wl}$ the threshold energy of white-light generation), the conical emission started to appear (Fig.4.3(b)).

---

[1]Throughout this work, the spectra recorded with a filter are normalized [0,1] to their peak intensity. The cut-off limits λ_{c_i} are the two wavelengths with a relative intensity at which below only noise is detected. The upper noise threshold is the highest measured intensity in the range, where the filter is supposed to block! (Wavelengths above specified filter cut-off ($\lambda >$@800nm,@785nm,@700nm), where the transmission is below $T < 10^{-5}$% 3.3). The threshold is individually defined in each spectrum (indicated by dashed line).

In chapter 2.1.2, it is discussed that this emission could result from diffraction of the white-light, or be related to a phase matched four wave-mixing [XYA93]. The red colored ring exhibits a divergence angle of $\theta = 5.5°$ on the inner and $\theta = 7.5°$ on the outer edge. No continuous connection between the center and the cone is visible. The central white spot is not a uniform disc anymore, but it splits up in several spots in different wavelengths located on different positions on the screen.

Upon further increasing the input pulse energy, the divergence angle of the central white spot and the cone increased as well (Fig.4.3(c)&4.4(a)). At $E_0 \approx 51.1\mu J$ ($2.88E_{wl}$), the conical emission exhibits a rainbow-like ring pattern, with a green and faint blue cone in addition to the red one Fig.(4.3(c)).

(a) $E_0 \approx 67.6\mu J$ ($3.82E_{wl}$)   (b) $E_0 \approx 202.8\mu J$ ($11.46E_{wl}$)   (c) $E_0 \approx 270.5\mu J$ ($15.28E_{wl}$)

Figure 4.4: *Part two of the white-light spot evolution projected onto a screen at different input pulse energies E_0 (E_{wl} measured energy threshold of continuum generation). No filter was used to block pump-laser.*

Here, it is also noticeable that the cones and the central part seem to exhibit some sort of a modulation pattern. This pattern results from an interference by the individual filaments and will be further discussed in section 4.6. The central part is now a circular white disk, no individual colored spots are visible anymore. The space between the conical emission and the white disk is filled with a light blueish colored ring.

At higher input energies (Fig.4.4(a)), the intensity distribution on the screen becomes more erratic as the beam breaks up in several more filaments (section 4.4). The central part now reaches the conical rings and starts to fill them. At $11.5E_{wl}$ the threshold ($E_0 = 202.8\mu J$), the intensity distribution (the spot on the screen) appears smoother (Fig.4.4(b)).

At a maximum input pulse energy of about $E_p \approx 270\mu J$ ($15.28E_{wl}$), the spectrally broadened pulse, appears on the screen as a solid white disk with blue rim and a total divergence angle $\theta = 15.6 \pm 1.2°$ (Fig.4.4(c)). At this point the red cone emission is partially covered by the central part of the beam. The corresponding spectrum is presented in Fig.(4.5).

A definite asymmetry in the spectrum and a stronger spectral broadening towards shorter wavelengths (higher frequencies), hence, a stronger anti-Stokes side can be found. The white-light continuum spreads from 459nm $\leq \lambda_{wl} \leq$ 1002nm (with normalized intensities $I \geq 0.8\%$).

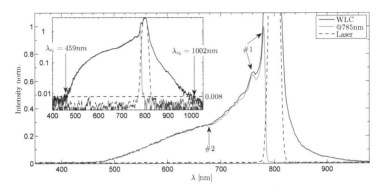

Figure 4.5: *Spectrum of the broadened pump laser beam at highest possible input pulse energy of $E_0 \approx 270.5\mu J$ (15.28E_{wl}). Visible are pump laser (dashed), White-light Continuum WLC (black solid) and @785nm filtered spectrum (red solid).*

It seems that the WLC spectrum exhibits some kind of modulation on the wavelengths near the residual laser peak (indicated #1, Fig.4.5). This may be explained by the interference of the newly generated frequencies, if their phase difference is $\Delta\phi_{12} = 2\pi$ for constructive interference or $\Delta\phi_{12} = \pi$ for destructive interference, as described in chapter 2. Another possible reason is an interference between the reflection of the white-light from the front and rear crystal barrier. The modulation has only been observed in some spectral measurements. However, its cause is of no importance to this work, and thus, not further investigated.

The @785nm filter transmits the entire anti-stokes side of the spectrum very well, due to its high transitivity. Small deviations from the unfiltered spectrum may result from temporal intensity variations during the measurement. The entire infrared part of the newly generated continuum is blocked and cannot be used for the surface structuring experiments.

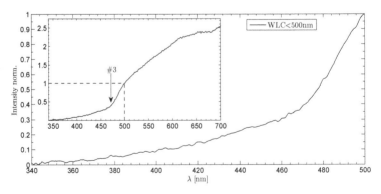

Figure 4.6: *White-light continuum spectrum in the range below $\lambda < 500nm$, recoded with the highest integration time setting of the spectrometer.*

The spectrum exhibits a steep edge near the wavelengths of the pump-laser and becomes flatter (smaller slope in narrow spectral intervals) at about $\lambda = 680$nm (indicated #2, Fig.4.5). The spectrum was recorded for a low integration time stetting of the spectrometer. However, below the lower cut-off at $\lambda_{c_1} \leq 459$nm of the spectrum only noise is detected.

A closer investigation with a higher integration time (higher number of processed pulses) setting reveals, that the white-light continuum actually spreads down to about $\lambda \approx 350$nm (Fig.4.6) at its lower wavelengths end. At about $\lambda \approx 470$nm (indicated #3, Fig.4.6) the spectrum exhibits another significant drop in slope and becomes even flatter.

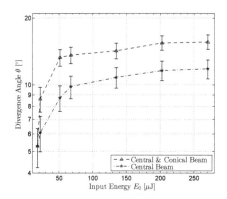

Figure 4.7: *Divergence angle θ of beam with conical emission (dashed) and central white-light part (dash-dotted) in dependance on the input pulse energy E_0.*

The dependence of the divergence angle θ on the input pulse energy E_0 is displayed in Fig.(4.7). Here, the dashed line represents the continuum beam including the conical emission. The beam spatially diverges fast from the $\theta_{wl} = 5.5°$ at the threshold of E_{wl}, up to $\theta = 13.3°$ at $2.9E_{wl}$ the threshold. For the lower input energy range, this means that by applying thrice the threshold pulse energy to the crystal, the divergence almost triples. Shortly after, the divergence goes into saturation.

The central white-light part of the beam shows the same progression, but exhibits at the highest input pulse energy a divergence value of $\theta = 11.8°$, which is $\Delta\theta = 1.5°$ less than for the complete beam. No statements can be made about larger input pulse energies, since the laser system only provided pulses up to 270μJ (at the position of the crystal). The error-bars were calculated via error propagation and are relatively large, because of the assumed random errors from the measurements. However, an overall tendency of the white-light beam divergence is observable.

4.2 Energy Conversion Efficiency & Continuum Behavior

In this section, the energy conversion efficiency, as well as the throughput efficiencies of filtered white-light (WL) continua (of the three short pass filters), is determined. The energy conversion efficiency provides an information of how much input pulse energy is converted to a WL continuum.

Under stepwise increase of the input pulse energy, the throughput energy of the WL continuum (WL + residual pump laser) was measured. The power-meter was placed in close proximity behind the rear end of the Al_2O_3 crystal. For the measurements of the filtered continua, the individual filter was placed between crystal and power meter. Thereby, the position of the filter and power-meter was not changed throughout the experiment.

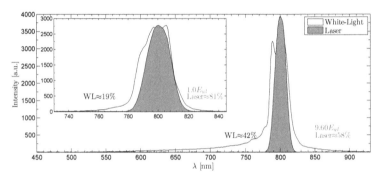

Figure 4.8: *White-light and pump laser area ratio, for $1.0E_{wl} = 17.7\mu J$ and $9.60E_{wl} = 170.02\mu J$ Continuum generation (Laser peak is normalized to WL peak). Spectra were recorded at the central part of the beam, where the pump-laser peak exhibits maximum intensity.*

In order to determine the efficiency, as well as the pulse energy of just the newly generated frequencies, a deconvolution of the WL data (WL without residual pump laser peak) from the measured data of the continuum (with residual pump laser peak) is performed.

At each corresponding input energy E_0, a spectrum of the continuum was recorded. All spectra were recorded under the exact same conditions, regarding integration time settings and position of the spectrometer to guarantee the comparability. From these spectra, the ratio of pump-laser (A_l) to continuum (A_c) was determined, by calculating the areas[2] under the curves:

$$\frac{A_l}{A_c} = x \ . \tag{4.1}$$

The ratio, which resembles the amount of pump-laser energy in the measured throughput energy of the continuum E_c, is subtracted from E_c, and, thus, yields the deconvolved WL pulse energy E_d:

---

[2]The areas under the curves were calculated from the N point data-sets via the *trapezoidal rule* (a numerical way of area approximation) given as $A = \int_a^b f(\lambda)d\lambda \approx (b-a)/2D \sum_{k=1}^{D}(f(\lambda_{k+1}) + f(\lambda_k))$, where a, b are the limits of integration/approximation and $D = N - 1$ the number of equally spaced data bins (here, 0.5nm bin).

$$E_d = E_c \cdot (1 - x) \ . \tag{4.2}$$

The area determination is shown in Fig.(4.8), for the two cases of minimum ($1.0E_{wl} = 17.7\mu$J) and maximum ($15.08E_{wl} = 267\mu$J) continuum generation at corresponding input pump energies.

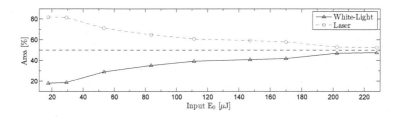

Figure 4.9: *Progression of white-light area ratio (without laser) and laser area ratio, for different input pump energies E_0.*

The white-light area percentage in the spectra increases for increasing pump energies and seems to go into saturation for high input pulse energies ($E_0 > 200\mu$J) (Fig.4.9). The energy conversion efficiency, as well as the efficiencies from the filtered continua, can be determined via the ratio of WL (or short pass filter) pulse energy E_d to input pulse energy E_0:

$$\text{Eff} = \frac{E_d}{E_0} \ . \tag{4.3}$$

In figure (4.10(a)) the behavior of obtained WL pulse energy versus the input pulse energy E_0 is displayed. In general, the WL pulse energy, as well as the pulse energies of the filtered continua, show an increasing progression.

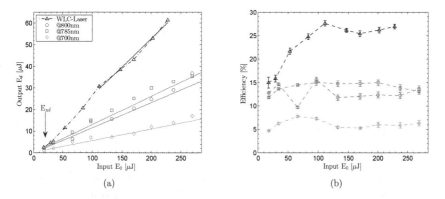

(a)   (b)

Figure 4.10: *(a) White-light energy vs. input pulse energy E_0 (deconvolved (Δ) from total energy and after transmission through filters). Linear fit of (@800nm, @785nm,@700nm filter) data points as well es for saturated WL data. (b) Energy conversion efficiency, as well as filtered continua efficiencies (Error-bars from estimated pulse energy errors, calculated via error propagation).*

The energy conversion efficiency of the deconvolved WL (Fig. 4.10(b)), increases from Eff=15.8% to maximum of Eff_{max} = 27.6%. In this input energy range the beam profiles on screen (Fig.4.3(a)-(c) Fig.4.4(a)), qualitatively evolve over proportionally in terms of divergence angle (Fig.4.7) and intensity distribution, upon small increase of pump energy.

Filter	Eff. [%]
WL	26.3±0.2
@800nm	12.8±0.5
@785nm	13.7±0.5
@700nm	5.9±0.4

Table 4.1: *Maximum white-light efficiency and fitted efficiencies of filtered continua (errors obtained from fit).*

At the peak, 27.6% of the applied input pulse energy are converted to new frequencies. At higher input energies, the efficiency drops to Eff=26.1% and seems to saturate. Here, the output energy E_d, is approximated by a linear fit with an average conversion efficiency of Eff_{ave}=26.3%. The drop and saturating efficiency might be related to a saturation in the continuum generation process, similarly reported in [TIm09], [ImF12]. The overall behavior of the WL conversion efficiency seems plausible, keeping in mind the nonlinear nature of of the optical process of continuum generation.

The usable pulse energies from the @800nm, @785nm and @700 filtered continua, exhibit also an increasing progression. However, even though the filters transmit the broader anti-stokes side of the spectrum, the contribution to the WL pulse energy becomes less at higher input energies E_0. For instance, the output energies from the @785nm filter, makes up only ≈ 57% of the deconvolved WL energy, at high input ($E_0 > 200\mu J$). The rest of the WL energy (≈ 43%) is contained in the infrared part, and high intensity wavelengths, close to the pump-laser peak (visible in Fig.4.8), which are not transmitted by the filter.

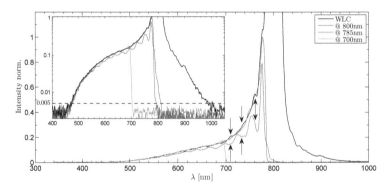

Figure 4.11: *White-light Continuum spectrum and transmitted spectra of the @800nm, @785nm & @700nm short pass filter, taken near the center of the beam. Arrows indicate destructive interference of WLC and @800nm modulation peaks.*

The efficiencies of all three filters show a slight increase of the efficiency in the same input energy region ($E_0 \approx 20-130\mu J$) as the WL. The @800nm filtered spectrum exhibits a fluctuation in efficiency at $E_0 = 65\mu J$. This may result from destructive interference of modulation peaks in the WL spectrum, with modulation valleys (Ch. 3.3) in filter spectrum (indicated arrows in Fig.4.11).

Thus, a high intensity/energy contribution is partially blocked and a lower output energy is measured.

The measured output energy of the filters (@800nm, @785nm, @700nm Fig.4.10(a)) is approximated by a linear fit. From the fit the average efficiency (slope of the fit) could obtained (table 4.1). The Eff=13.7% efficiency from the @785nm filtered spectrum, is slightly higher, than Eff=12.8% from the @800nm spectrum, since the @785nm filter (section 3.3), exhibits a higher average transmission, despite lower cut-off. Thus, with the @785nm filter, a maximum output pulse energy of $E_d = 35.5\mu J$ could be obtained by pumping the crystal with $E_{max} = 269\mu J$ laser pulses. Only $\approx 6\%$ of the pump energy is converted to wavelength below $\lambda < 700$nm.

A comparison between the transmitted spectra of the white-light continuum is displayed in Fig.(4.11). The @700nm spectrum exhibits a flatter progression, since it does not contain the high intensity wavelengths (steeper slope) near the $\lambda_{laser} = 800nm$ pump-laser peak, compared to the @800nm & @785nm spectra.

The filters block the pump-laser peak and transmit the lower end of the white-light spectrum well, which makes them suitable for the investigation of LIPSS with a white-light continuum.

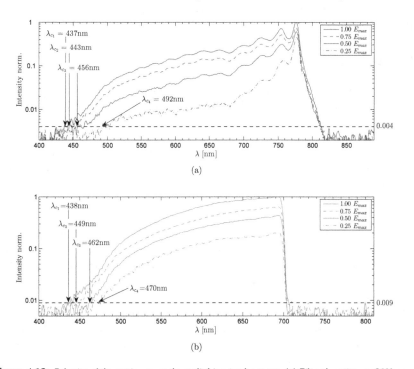

Figure 4.12: *Behavior of the continuum on the applied input pulse energy.(a) Filtered continuum @800nm cut-off The modulation arises from the dielectric multi-layer nature of the filter (Ch. 3.3), (b) @700nm cut-off. Spectra in 25% steps to maximum input pulse energy E_{max}.*

In figure (4.12) the behavior of the @800nm and @700nm filtered continua upon applied input pulse energy is demonstrated. As discussed in chapter 2.1.2, the spectral broadening of the pump-laser is intensity dependent, thus for lower pump inputs the spectra should cut-off at loger wavelengths. The measured spectra are presented in quarter steps of the maximum input energy E_{max}. At maximum pump energy E_{max}, the @800nm continuum, displayed in Fig.(4.12(a)), spreads down to $\lambda_{c_1} = 437$nm (normalized intensity at $I = 0.4\%$, upper noise threshold limit).

By decreasing the input pulse energy the lower end cut-off, takes place at longer wavelengths. At an input of $0.75E_{max}$ the cut-off moved to $\lambda_{c_2} = 443$nm, at $0.50E_{max}$ to $\lambda_{c_3} = 456$nm and at $0.25E_{max}$ to $\lambda_{c_4} = 492$mm.

With the @700nm cut-off filter (4.12(b)), wavelengths down to $\lambda_{c_1} = 438$nm (normalized intensity at $I = 0.9\%$, upper noise threshold limit), are detected for the maximum input E_{max} and low integration time setting of spectrometer[3]. At $0.75E_{max}$ the cut-off moved to $\lambda_{c_2} = 449$nm, at $0.50E_{max}$ to $\lambda_{c_3} = 462$mm and at $0.25E_{max}$ to $\lambda_{c_2} = 470$nm.

This cut-off behavior of the transmitted continua, shows that only at high input pulse energies, the broadest spectrum, in terms of relative intensities above the defined upper noise threshold limit can be obtained.

4.3 White-Light Polarization

The polarization of the white-light is of essential importance in the experiments of LIPSS formation, since it is one of the control parameters guiding the direction in which the ripples grow. Thus, the polarization of the white-light continuum was investigated with a *Thorlabs - LPVIS100* linear polarizer, which acted as an analyzer. It covered the spectral range from 550nm$\leq \lambda \leq$1650nm and was placed behind the 800nm filter short pass filter (used in this set up). Subsequently, the throughput pulse energy was measured for different rotation angles α of the analyzer.

The incident pump laser from the Spitfire amplifier is linearly polarized in horizontal direction (with respect to the work table). At the rotation angle of $\alpha = 0°$, the throughput direction is parallel to the incident light polarization and at $\alpha = 90°$ perpendicular. The measurement in Fig.(4.13(a)) was carried out with an input pulse energy of $E_0 = 170\mu J$. A maximum of the measured power for rotation angle of $\alpha = 0°$ and a minimum at $\alpha = 90°$, can be noticed. This means that in general the white-light is polarized and preserves the polarization direction of the pump-laser.

To quantify the polarization of the white-light beam the extinction ratio ER has been used. The extinction ratio is defined as the ratio between the transmitted power when the analyzer is in perpendicular position ($P_\perp$, $\alpha = 90°$) and when it is in parallel position ($P_\parallel$, $\alpha = 0°$):

$$ER = \frac{P_\perp}{P_\parallel} .\qquad(4.4)$$

---

[3]Here, a low integration time setting of the spectrometer is used, because at a higher integration time the spectrometer went into saturation.

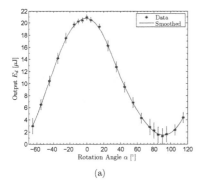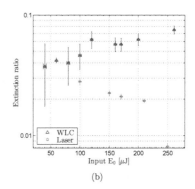

(a)   (b)

Figure 4.13: *(a) Polarization curve for different rotation angles α of the analyzer ($E_0 = 170\mu J$, 800nm filter). The blue line represents smoothed data points, while the error-bars are estimated errors depending on the variation of the measured power value. (b) Extinction ratio of the WLC and pump-laser beam for different input pulse energies.*

For the measurement in Fig.(4.13(a)) an extinction ratio of $ER_{wl} = 0.057\pm0.007$ was obtained. At this input pulse energy, the incident laser has an extinction ratio of $ER_{laser} = 0.021 \pm 0.001$. The plot of the extinction ratio versus the input pulse energy, displayed in Fig (4.13(b)), shows a clear tendency of laser (Spitfire output beam) and transmitted white-light continuum (full setup).

With increasing input energy the extinction ratio of the laser decreases. This is related to the extinction degree of the thin-film polarizer, used in the energy attenuation part of the setup (Fig.3.2). The laser beam leaving the polarizer still exhibits off-axis polarization components which contribute to the polarization in the output beam. This contribution is stronger for low attenuated energies.

The white-light beam, however, exhibits the opposite case. With increasing input energy the extinction ratio degrades by a factor of 5 and light leaving the crystal becomes progressively depolarized, hence the white-light cannot be extinct that well anymore. At the highest input energy the white-light ER is one order of magnitude higher,than the ER of the laser.

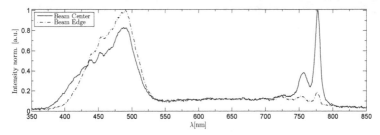

Figure 4.14: *Spectra of the extinct white-light beam taken at the central and outer part of the beam.The peaks of both spectra are normalized [0,1], and thus can be compared. The residual peaks on the lower end of the spectrum, result from the finite range of the analyzer [550nm,1650nm].*

In [DRM06] it has been suggested that plasma effects in the bulk material, due to multi photon excitation, are responsible for a depolarization at larger input energies. A spectral measurement at the maximum energy, of the beam's cross section revealed, that the depolarization tends to take place in the central part. The spectrum in the beam center, displayed in Fig.(4.14), exhibits intensity peaks, which are not extinct, near the λ_{laser}=800nm pump-laser. Hence, their direction of polarization differs from the rest of the beam. At the edge of the beam (dashed line) these peaks have a significantly lower intensity, while rest of the spectrum shows intensities similar to the beam center.

The plateau in the range from 800nm down to 550nm suggests, that the light here is depolarized as well. So it is to say that in majority linear polarization of the pump-laser is preserved during the continuum generation in the Al_2O_3 crystal.

4.4 Spatial Properties

As described in chapter 2, the WL beam breaks up in multiple filaments in transverse direction during its propagation through the Al_2O_3 crystal, if the power of the pump-laser exceeds the critical power for self-focusing. Each of the filaments then undergoes their own self-focusing and self-phase modulation. This behavior of the white-light continuums spatial distribution, has been monitored with the CCD camera for different input pump energies (Fig.4.15, 4.16).

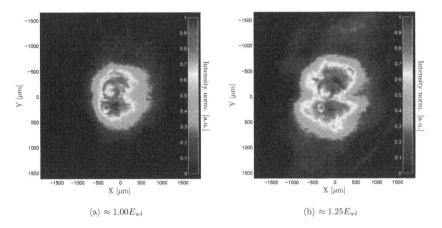

(a) $\approx 1.00 E_{wl}$ (b) $\approx 1.25 E_{wl}$

Figure 4.15: *(a) CCD Images of the WL beam transverse intensity profile at $\approx 1.00 E_{wl}$, (b) at $\approx 1.25 E_{wl}$. CCD images were taken about 10cm behind the crystal. An explanation on how the scale sizes in [μm] was obtained, is given in Ch.5.1. Images are false-colored (color-scale) and normalized [0,1]!*

At roughly onset threshold of continuum generation, the transverse intensity distribution of the beam shows a single coherent spot, with the maximum intensity located in its center (Fig 4.15(a)). For $\approx 1.25 E_{wl}$ (Fig.4.15(b)), the central part consists of multiple spots.

At this point conical emission is visible in the farfield (on-screen profiles) in Fig (4.3(a)). In the near-field, the intensity spots in the periphery from the center, may cause the emission. However, in [BIC96] it is suggested, that the conical emission originates from inside the filaments. The intensity distribution shows, that the beam split up vertically along his propagation axis, in two filaments, each of which carrying maximum intensity in its center.

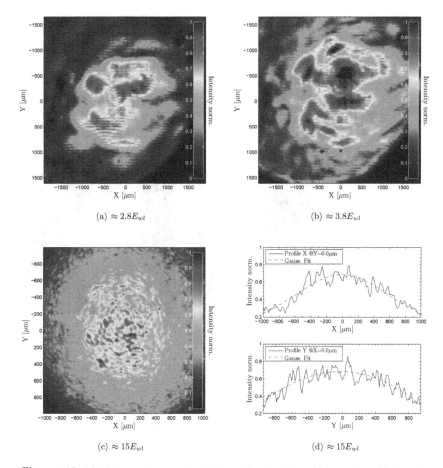

(a) $\approx 2.8 E_{wl}$ (b) $\approx 3.8 E_{wl}$

(c) $\approx 15 E_{wl}$ (d) $\approx 15 E_{wl}$

Figure 4.16: *(a) CCD image of transverse WL beam profile at $\approx 2.8 E_{wl}$, (b) at $\approx 3.8 E_{wl}$. (c) At $\approx 15 E_{wl}$ and (d) corresponding intensity profiles in X&Y-direction. Images are false-colored (color-scale) and normalized [0,1]!*

Upon further increasing the input pulse energy to $\approx 2.8 E_{wl}$ (Fig.4.16(a)), the beam splits up in several more filaments. Thereby, filaments in the periphery from the center exhibit higher intensities as well.

At $\approx 3.8 E_{wl}$ (Fig.4.16(b)) the white central part starts to cover the conical emission, (recall Fig.(4.4(a))). Here, the intensity profile consists of many individual filaments, with an accumulation near the central axis of the beam.

The filament spots were fluctuating in intensity and position. Thus, the images only represent a brief moment of the beam intensity variation.

Figure (4.16(c)) shows the transverse intensity profile at the maximum possible input pulse energy, at $\approx 15 E_{wl}$. Because of the large divergence angle at this energy, the camera had to be moved closer to the crystal to capture the whole profile. Hence, the axes are not comparable with the previous images in Fig.(4.15(a)-4.16(b)). The beam splits up in very many filaments. The highest intensities are located near the beam axis, with normalized intensity variations up to $\Delta I/I \approx 50\%$. Corresponding intensity profiles in x&y-direction are displayed on the plots in Fig.(4.16(d)). The profiles show the intensity peaks resulting from the breakup in many filaments. The envelope reflects a Gaussian distribution with a large width (indicated fit).

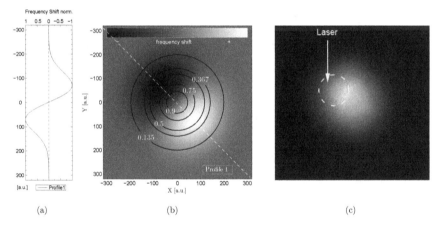

(a)   (b)   (c)

Figure 4.17: *SPM in spatial domain. (a) Cross section profile indicated in b, (b) Isohypses of Gaussian transverse beam profile at $1/e^2$, $1/e$, 0.5, 0.75 and 0.90 of I_{max}, grey-scale shows derivative (frequency shift Eq.4.5). (c) White-light beam profile on a screen, at $1.0 E_{wl}$ onset of continuum generation.*

Responsible for the spectral broadening of the pulse is the self-phase modulation. As mentioned in chapter 2.1.2, SPM can also take place in the spatial domain, which then results in spectral differences across the transverse beam profile. Since the spatial distribution of the provided laser pulses is Gaussian, the central part exhibits the highest intensity, compared to the wings (Isohypses, lines of equal intensity, in Fig.4.17). The frequency broadening (SPM), depends on the slope of the intensity,

$$\Delta\omega \propto -\left[\frac{\partial I(x,y)}{\partial x} + \frac{\partial I(x,y)}{\partial y}\right] . \tag{4.5}$$

Figure (4.17(b)) shows a contour plot (isohypses at specific values) of a Gaussian beam profile. The grey-scale indicates the sum of the derivative in both spatial dimensions, hence, the location of the resulting frequency shift of the pulse[4]. In the vicinity of the peak intensity, the slope is virtually zero, which means almost no frequency shift (Fig4.17(b)). Thus, in the central part wavelengths close to the pump laser wavelength (λ_{laser}) are located.

In positive direction from the center (Fig.4.17(a)), the slope is negative and the laser frequency is shifted towards higher frequencies, while in opposite direction a shift to lower frequencies takes place. Consequently the beam exhibits a distinct spatial distribution of frequencies. One needs to keep in mind that with this simple approximation, the resulting spectrum would be symmetric and furthermore exhibit quasi periodic oscillations in space (similar to a ring shaped diffraction pattern), as discussed in Fig. (2.4). Nevertheless, this spatial distribution of frequencies could be observed at the onset of continuum generation (Fig.4.17(c)).

A clear blue broadening in one direction, as well as a shift to lower frequencies in the opposite direction, similar to the frequency shift grey-scale in Fig.(4.17(b)) is visible. The arrow indicates the location of the pump laser spot.

At high input pulse energies the various filaments inside the crystal undergo their own self-phase modulation. Thereby, wavelengths at the individual spatial positions of each filament superimpose on the screen and the beams central part appears as a white disk.

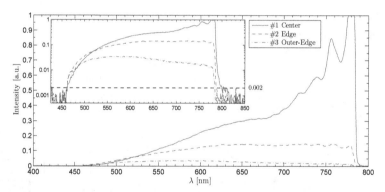

Figure 4.18: *Filtered white-light spectra (@785nm) taken in the central part (#1) in the periphery (#2) and outer region (#3) of the beam. Corresponding positions are displayed in Fig.(4.19). Spectra recorded with a low integration time setting of the spectrometer*

The spectral distribution still exhibits as strong spatial dependence, demonstrated in for $E_0 \approx 270\mu J$ ($15.2E_{wl}$) pump energy.

Figure (4.18) shows the filtered spectrum of the WL-continuum for the @785nm filter at three different position on the beams transverse profile. The spectra were recorded in the center of the white-light-disk (#1), in the periphery from the center (#2) and outer region of the beam (#3).

---

[4]See also indicated profile Fig.(4.17(a)), comparable with SPM in temporal domain Fig.(2.3).

Figure 4.19: *Spatial positions of the three spectra in Fig.(4.18).*

During the measurements the spectrometer was covered with scattering paper, thus, light around the marked position is also scattered into the spectrometer resulting, for instance, in the detection of higher wavelengths at position #3.

From Fig.(4.18), one can clearly see a major difference the three spectra. The spectrum from the center of the beam exhibits a very steep slope near the pump-laser wavelength and becomes progressively flatter (moderate slope in narrow spectral intervals) at about $\lambda = 700$nm. Here, the modulation peaks near the 785nm cut-off are not caused by the filter.

In the periphery from the center, the continuum exhibits the flattest possible spectrum with an almost constant intensity from $\lambda = 785$nm down to $\lambda = 600$nm. It thereby has only $\sim 15\%$ the intensity of the peak intensity in the beam's center. Wavelengths below $\lambda < 500$nm show here a higher intensity than in the central part of the beam. The spectrum from the outer region exhibits highest intensity at shorter wavelengths, with a maximum at around $\lambda \sim 550$nm. One can expect such a spectrum at this position from the green and blue rim of the beam profile (#3, Fig.4.19). However, here the light is so faint (only $\sim 3\%$ of the peak from the beam's center), that it does not contribute significantly to the WL pulse energy (fluence).

4.5 Long Term Stability

Another important investigation concerns the stability of the continuum over longer time periods. During the generation of LIPSS, some targets will be exposed to a high irradiation dose. This means, they will be irradiated for a couple of minutes with pulses at a constant fluence. With a repetition rate of $f_{rep} = 1$kHz, this makes more that 10^6 WL pulses.

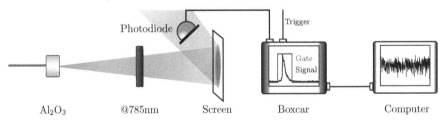

Figure 4.20: *Sketch (not to scale!) of the setup for measuring the white-light continuum long term stability. Setup consist of a photodiode for visible range, Gated Integrator and personal Computer for data readout and averaging. The Continuum was generated by focusing the laser beam with the +150mm lens into the crystal (not displayed here).*

The stability of the white-light continuum was monitored with a fast photodiode, specified for the visible part of the electromagnetic spectrum. A sketch of the setup for this measurement is shown in Fig.(4.20). The generated continuum was projected onto a screen and the pump-laser blocked behind the crystal by the @785nm filter.

The photodiode was placed in the scattered white-light from the screen and further dimmed by neutral density filters.

In order to measure such a high amount of events over time, the signal from the diode was integrated and averaged using the *Stanford Research Systems Model SR250 Gated Integrator*. The integrator places a square gate function over the signal (indicated in Fig.(4.20)), which is triggered externally from the main laser system. The signal from the photodiode is, with $t_{diode} \approx 800 \times 10^{-9}$s, fast enough to discriminate every pulse. Thus, the width of the gate was set to t_{diode}. Subsequently, the integrator averages over 200 signal samples.

The long term stability was measured over a period of 1200s (20min) for an input pulse energy of $E_0 = 285.2\mu$J, displayed in Fig.(4.21).

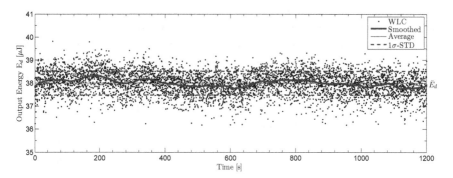

Figure 4.21: *White-light continuum stability measurement over a period of 1200s (20min).*

The pulse energy of the white-light, measured with the power-meter, was at $\bar{E}_d = 38\mu$J ($\bar{P}_d$=38mW). Because the power-meter is based on the absorption of heat it is not fast enough do detect pulse to pulse energy variations. Which means that the value $\bar{E}_d$ is already a time average over all the pulses hitting the absorber. Thus, the average intensity from the diode (calculated from all data points), can be scaled with at $\bar{E}_d = 38\mu$J.

Over the measured 20min. time period, the white-light signal appears stable and does not break up or drastically decrease over magnitudes of tens of μJ. The largest found deviation from the average is $\Delta \bar{E}_d = 1.76\mu$J. Each data point in the plot thereby, represents the average of 200 white-light pulses. The dashed lines represent the standard deviation, at a value of $(\pm 1\sigma - \bar{E}_d) = 0.52\mu$J with respect to the average. This corresponds to a 2.73% pulse to pulse variation boundary and is thereby, close to the specified 3% pulse to pulse stability of the pump-laser system (Ch. 3).

Variations in the pulse energy can be seen better in the smoothed signal, which is calculated via the Savitzky-Golay [SGo64] data smoothing method. The smoothed signal stays within the standard deviation bounds. Most of the shot to shot fluctuations, however, arise from the fluctuations of the pump laser system. Thus, the stability of the white-light is in good agreement with the pump laser specification. It is to say that the generated white-light continuum exhibits a long term stability which makes it suitable for measurements over larger periods.

4.6 Spatial Coherence

In wave-optics the common usage of the term coherence is always linked to the ability of light to produce an interference pattern. If the light is mutually coherent in the spatial domain, this means that two points on an extended source have a constant phase relation and can interfere constructively, which produces an interference pattern. If these two points radiate independently from each other, the emitted light waves do not exhibit a constant phase relation and the waves cannot interfere, hence the light is mutually incoherent.

The spatial coherence of light waves, thus depends on the physical size of the source and determines how far two points can be separated in a plane transverse to the direction of propagation of the light and still be correlated in phase [Sha06].

The coherence of light in the temporal domain, however, determines how far in distance two points along the direction of wave propagation can be and still possess a definite phase relationship. Temporal coherence would be directly related to the spectral bandwidth of the light source. Since the spectral bandwidth spreads from about 350nm up to 1000nm, one can assume that the temporal coherence of the white-light is very small. However, it is of minor interest for this work and will thus, be not further discussed here.

The coherence of the white-light continuum has been investigated in the spatial domain with a double pinhole experimental setup (Fig.4.22).

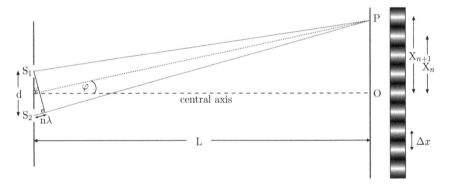

Figure 4.22: *Schematic setup of double pinhole experiment and derivation of fringe separation.*

This kind of setup is sometimes referred to as *Young's Double Slit Experiment*, since it was first performed at the beginning of the 19*th* century by the English polymath Thomas Young. If the double slit is irradiated by a coherent light source an interference pattern will be visible on the screen. If light emitted from the source has a reduced coherence or is virtually incoherent, a reduced contrast or visibility V, in the interference fringes or no interference pattern at all will be visible, correspondingly. At the distance L to the pinhole plane a screen is mounted. This screen is in the experiments replaced by the monochrome CDD camera.

The pinholes have a separation of d. At the point $O(\vec{r})$, the direct opposite mid-point between $S_1(\vec{r}_1)$ and $S_2(\vec{r}_2)$, the path difference between waves $S_2O - S_1O$ is zero.

Thus, constructive interference occurs and the central fringe or maximum is bright. Suppose $P(\vec{r})$ is the position of the n-*th* order bright fringe. The optical path difference between the two sources $S_1(\vec{r_1})$ and $S_2(\vec{r_2})$ must differ by an integer number of wavelengths:

$$S_2 P - S_1 P = n\lambda \ . \tag{4.6}$$

For a distance L much larger than the pinhole separation d, the path difference can be approximated by $d \cdot \sin\varphi = n \cdot \lambda$. The distance of the n-*th* bright fringe from the central axis is given as $d \cdot \tan\varphi = X_n$. For small angles φ, one can approximate $\sin\varphi \approx \tan\varphi \approx \varphi$, hence $X_n = n\lambda L/a$. Thus, the separation between adjacent fringes (i.e. fringe separation) is given by the equation:

$$\Delta x = X_{n+1} - X_n = \lambda \cdot \frac{L}{d} \ . \tag{4.7}$$

The contrast between dark and bright fringes in the interference pattern is also referred to as visibility $V(\vec{r})$. It can be derived in the same way. The total instantaneous electric field $\vec{E}(\vec{r},t)$ at point P is the sum of the two sources $\vec{E}(\vec{r},t) = \vec{E}_1(\vec{r_1},t) + \vec{E}_2(\vec{r_2},t)$. The square of the total field is proportional to the energy (Poynting) flux $S(\vec{r},t)$:

$$S \propto E^2 = \left(\vec{E}_1 + \vec{E}_2\right)^2 = E_1^2 + E_2^2 + 2\vec{E}_1\vec{E}_2 \ . \tag{4.8}$$

With that the intensity of the light at point P may be obtained from the time average of S:

$$I(\vec{r},t) = \langle S(\vec{r},t)\rangle \propto \left\langle E_1^2(\vec{r_1},t)\right\rangle + \left\langle E_2^2(\vec{r_2},t)\right\rangle + 2\left\langle \vec{E}_1(\vec{r_1},t)\vec{E}_2(\vec{r_2},t)\right\rangle \Re e\left[\gamma(\vec{r_1},\vec{r_2})\right] \ . \tag{4.9}$$

The cross term $2\left\langle \vec{E}_1(\vec{r_1},t)\vec{E}_2(\vec{r_2},t)\right\rangle \Re e\left[\gamma(\vec{r_1},\vec{r_2})\right]$ represents the correlation between the two light waves. The term $\Re e\left[\gamma(\vec{r_1},\vec{r_2})\right]$ is the real part of the complex degree of coherence [Sah10]. The modulus of this function lies in the range of $0 \leq |\gamma(\vec{r_1},\vec{r_2})| \leq 1$ and has a maximum value at the origin. It thereby equals the contrast of the interference fringes or visibility:

$$V(\vec{r}) = |\gamma(\vec{r_1},\vec{r_2})| = \begin{cases} = 1 & \text{completely coherent} \\ < 1 & \text{partially coherent} \\ = 0 & \text{incoherent superposition} \end{cases} \tag{4.10}$$

In the first case, light waves from $S_1(\vec{r_1})$ and $S_2(\vec{r_2})$ are completely coherent. In the second case the light waves are partially coherent and the source operates at a reduced degree of coherence. If the source at the separation of d, radiates independently, no interference pattern will be visible and the cross term vanishes. Thus, the intensity on the screen is simply the sum of the two individual intensities $I_{inc} = I_1 + I_2$. The source is then said to be an incoherent superposition of light waves. For coherent sources, the cross term is non-zero. In fact, for constructive interference $(n\lambda)$, the maximum intensity is given by:

$$I_{max} = I_1 + I_2 + 2\sqrt{I_1 I_2}\ |\gamma(\vec{r_1},\vec{r_2})| \ , \tag{4.11}$$

and for destructive interference the intensity takes on a minimum value of:

$$I_{max} = I_1 + I_2 - 2\sqrt{I_1 I_2}\ |\gamma(\vec{r}_1, \vec{r}_2)|\ . \tag{4.12}$$

The visibility of the fringes $V(\vec{r})$, at point $P(\vec{r})$ is estimated in terms of the intensity of the two beams as:

$$V(\vec{r}) = \frac{I_{max} - I_{min}}{I_{max} + I_{min}}\ . \tag{4.13}$$

$V(\vec{r})$ can be calculated from the maximum and minimum intensity value in the vicinity of the central axis, which means zero path difference between the light waves from $S_1(\vec{r}_1)$ and $S_2(\vec{r}_2)$ [Sah10].

For the investigation of the white-light spatial coherence as well as of the pump-laser beam, 4 different double-pinhole arrays, with a pinhole separation of $d_1 = 350\mu m$, $d_2 = 600\mu m$, $d_3 = 1.1mm$, and $d_4 = 2.6mm$ have been manufactured. By focusing the pump-laser beam with a +300mm-lens onto a 0.25mm thin aluminum plate the laser then drilled the pinholes into the plate. From the investigation of the spatial intensity distribution in Appendix B, it is known that the diameter of the waist in the focus are $\bar{d}_{0x,300} \approx 90.22\mu m$ and $\bar{d}_{0y,300} \approx 81.98\mu m$ at the $1/e^2$ relative intensity value.

Figure (4.23) shows an SEM image of the pinhole array with $d_1 = 350\mu m$ hole separation and good correspondence to the determined waist values. The manufacturing process was limited by the precision of the translation stages used for the separation of the pinholes.

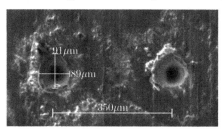

Figure 4.23: *SEM image of the manufactured double pinhole array with $d_1 = 350\mu m$ hole separation. The diameters of 91μm and 89μm correspond to the two axes of the laser beam (Fig.3.1)*

4.6.1 Filament Interference

During the investigation of the spatial intensity profile in section 4.4, it was found that at onset of continuum generation ($E_{wl} = 17.7\mu J$) a single filament formed inside the Al_2O_3 crystal (Fig.4.15(a)). Upon further increasing the input pulse energy the beam breaks up in several filaments, distributed transverse to its propagation axis (Fig.4.16). Now, while watching the screen, one could observe the formation of a modulated light pattern, already without a double pinhole aperture, when the input energy just slightly exceeds E_{wl}. This pattern has been reported to be an interference pattern from two or more filaments inside the crystal [WIt01], [BHä00].

The white-light emitted from the filaments has at this point a fixed phase relation, and thus, can yield the narrow interference pattern. Hence, it is to say that a high degree of coherence is preserved for lower input energies. Some of the interference patterns of the white-light are displayed in Fig.(4.24) for different input pulse energies near the threshold E_{wl}.

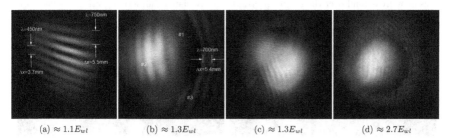

(a) $\approx 1.1 E_{wl}$ (b) $\approx 1.3 E_{wl}$ (c) $\approx 1.3 E_{wl}$ (d) $\approx 2.7 E_{wl}$

Figure 4.24: *Interference patterns of two or more white-light beam filaments displayed on a screen at different input pulse energies. (b) and (c) recorded at the same pump energy, yet slightly different z-positions of the crystal.*

The pattern changes for a constant input pulse energy, if the Al_2O_3 crystal is slightly moved in longitudinal direction (z-scan) with respect to the laser beam. A z-scan results in different self-focusing dynamics and therefore different filament distributions. The first pattern in Fig. (4.24(a)) appeared at $\approx 1.1 E_{wl}$. Here, no conical emission is observed. Since the fringes extended horizontally across the screen, this means that the two filaments generating this pattern are arranged vertically. One can see the spatial dependence of the wavelength distribution. The pump-laser was blocked with the @785nm filter, however, the pattern was also visible without the filter.

At about $\approx 1.3 E_{wl}$ (Fig.4.24(b)), the conical emission was clearly visible. Here, three different interference patterns can be identified (#1-3). By slightly moving the crystal more in z-direction the pattern in Fig. (4.24(c)) was obtained. Here, one can also see a superposition of three interference patterns, two in the center with different orientation and one irregular pattern in the conical emission. Upon higher input pulse energies ($> 2.0 E_{wl}$) (Fig.4.24(d)), the interference patterns became more irregular since the light of many filaments interfered with each other. Above pump energies is $> 3.0 E_{wl}$, no filament interference patterns were visible anymore.

From the fringe spacing of the interference patterns in Fig.(4.24(a))& (4.24(b)), one can derive the spacing of the two filaments generating the pattern. In figure (4.24(a)) the spacing between two bright fringes in the upper right part is about $\Delta x = 5.5 \pm 0.1$mm. In this part of the pattern the longer wavelengths are located, thus, lets assume that the $\lambda = 750 \pm 50$nm wave interfered with each other. Hence, for a distance $L = 12 \pm 0.3$cm of the screen to the center of the crystal, the traverse separation of the filaments inside the crystal, is after solving Eq.(4.7) for $d_{750nm} = 16.4 \pm 1.4 \mu$m. On the left-hand side the shorter wavelengths interfere. Here, the fringe spacing is $\Delta x = 3.7 \pm 0.1$, thus, for a wavelength of $\lambda = 450 \pm 50$nm, the filament separation is $d_{450nm} = 14.6 \pm 2.0 \mu$m. This gives a good approximation for the filament separation, generating this pattern.

The interference pattern of the conical emission, in Fig.(4.24(b)), shows a fringe spacing of $\Delta x = 5.4 \pm 0.1$. For interfering waves of $\lambda = 700 \pm 50$nm, the filaments are separated by $d_{700nm} = 15.5 \pm 1.4\mu$m.

In figure (4.25) it is demonstrated, that the interference patterns of the two above discussed cases actually extend over the entire spectral range. A planar diffraction grating has been used

Figure 4.25: *Spectrally dispersed white-light finges of the interference patterns in Fig.(4.24(a))& (4.24(b)).*

for angular dispersion of the white-light continuum. However, one can see, that the contrast of the dark and bright fringes fades as the wavelength becomes smaller (blue). This is a sign that the spatial coherence is reduced. Such measurements of the visibility, will now be discussed for high pump energies, using the double pinhole arrays.

4.6.2 Pump-Laser

At first the spatial coherence of the pump-laser was investigated. Therefore, the double pinhole setup was placed ∼1m from the outlet of the Spitfire amplifier. The laser spot was unfocused at the pinhole array and thereby exhibited a diameter of $d_{laser} \approx 5.5$mm.

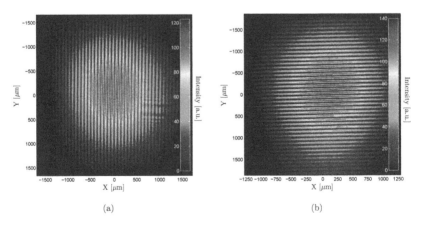

(a)   (b)

Figure 4.26: *Raw intensity images using the $d_1 = 350\mu$m double pinhole array. (a) Interference pattern generated by pinholes oriented along the horizontal axis, and (b) along the vertical axis of the pump-laser beam. Images are false-colored (color-scale) with arbitrary unit intensity scale!*

Subsequently, the interference pattern was dimmed with a neutral density filter, so the CCD camera would not go in saturation. The light emitted from a laser is supposed to be mutually coherent in time and space, since a definite phase relation between the light waves is a fundamental property of the laser. The phase relation originates from the stimulated emission In figure (4.26), the interference patterns of the pump-laser, for horizontal (4.26(a)) and vertical (4.26(b)) pinhole orientation are displayed. The patterns were produced by the pinholes with $d_1 = 350\mu m$ separation. Such raw-intensity images of the interference patterns, were now used to extract the intensity profiles across the center of the pattern. The extracted profiles, displayed in Fig.(4.27) for all 4 pinhole separations, are always perpendicular to the fringe pattern.

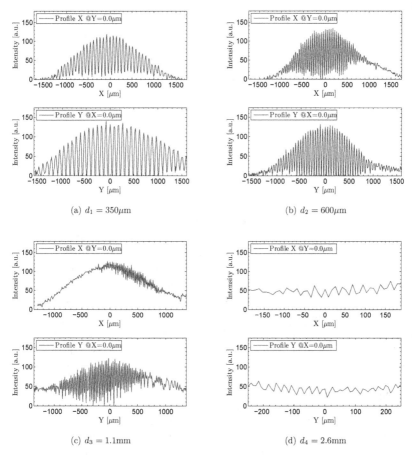

Figure 4.27: *Intensity cross-section profiles of pinholes oriented along the horizontal axis and along the vertical axis of the pump-laser beam.*

From the intensity cross-section profiles one can now calculate the visibility of the fringes V with Eq.(4.13), and thus, get the information about the spatial coherence of the source. This has been done in the following way.

The minimum and maximum intensity values, I_{min} and I_{max}, were measured for the central fringe (0μm fringe separation), as well as for one additional fringe located on the right-hand next to the center and one located on the left-hand to the center. The reason is that the intensity of the camera pixels also seemed to fluctuate, thus the mean value of $\bar{I}_{max}$ & $\bar{I}_{min}$ could then be used for the calculation of V.

The random errors from $\Delta\bar{I}_{max}$ & $\Delta\bar{I}_{min}$ were obtained by taking the largest difference of a measured value to the mean value, $\Delta\bar{I}_{max} = I_{max} - \bar{I}_{max}$ The complete visibility error ΔV, was then calculated via error propagation. The visibility V versus the pinhole separation d is shown in Fig.(4.28).

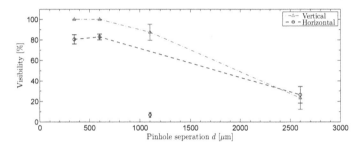

Figure 4.28: *Plot of the fringe visibility (%) as a function of pinhole separation, for horizontal laser beam axis (pinhole orientation)(dash-dotted) and vertical axis (dashed).*

For hole separations up to $d = 600\mu$m, the visibility is $V_y = 100\%$, on the vertical transverse axis of the pump-laser beam. This means that the modulus of the complex degree of coherence $|\gamma(\vec{r}_1, \vec{r}_2)| = 1$, and thus, the laser is vertically completely coherent in this hole separation range.

Horizontally, the beam only exhibits a visibility of $V_x = 83\%$. For larger hole separations the visibility of both axis falls to an almost identical value of $V_{x\&y} = 23\%$. The visibility value at $d = 1100\mu$m for horizontal orientation has been excluded from the line intersections, since an error during the measurement may have occurred.

Overall, the beam shows a higher spatial coherence on its vertical axis. If one recalls the properties of the beam from chapter 3, the beam exhibits an elliptical shape with a longer vertical axis.

4.6.3 White-Light

The visibility of the interference fringes, and thus, the spatial coherence of the white-light continuum has been investigated in the same way. The continuum was generated in the Al$_2$O$_3$ crystal with the highest possible input pulse energy $E_0 \approx 270\mu$J. Subsequently, the $\lambda_{laser} = 800nm$ pump-laser was blocked by the @785nm cut-off filter. Due to the large divergence, the beam was collimated by a +100mm-lens in order to propagate with a constant diameter of $d_{wl} \approx 2$cm. The generated interference pattern was attenuated with the neutral density filter and recorded by the CCD camera. A sketch of this arrangement is displayed in Fig.(4.29).

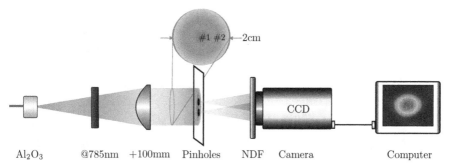

Al$_2$O$_3$ @785nm   +100mm   Pinholes   NDF   Camera   Computer

Figure 4.29: *Sketch (not to scale!) of the setup for WLC spatial coherence measurements. The continuum was generated by focusing the laser beam with the +150mm lens into the crystal (not displayed here). A neutral density filter (NDF) was mounted right in front of the CCD camera, so no stray-light could enter the camera. Indications #1, #2 correspond to the positions of the measurements at the beam center and outer part (conical emission).*

Figure (4.30) shows the raw intensity images of the interference pattern produced by the double pinhole array with $d_1 = 350\mu$m pinhole separation. Already a difference in the fringe

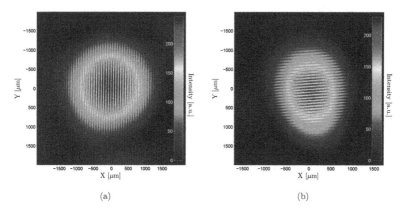

(a)   (b)

Figure 4.30: *Raw intensity interference pattern images, produced with $d_1 = 350\mu m$ pinhole separation at max. pump energy. Images are false-colored (color-scale) with arbitrary unit intensity scale!*

contrast compared to the images in Fig.(4.26) from the pump-laser can be noticed. Here, the intensity of the dark fringes is not zero, which is a sign for reduced spatial coherence. The spatial coherence measurements of the white-light were carried out for the central part (#1 indicated in Fig.4.31) as well as for the outer part (with conical emission) of the beam at a location of about ≈ 5mm from the center (#2 indicated in Fig.4.32).

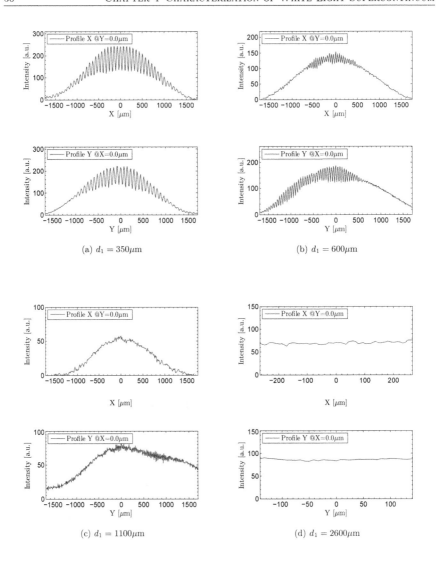

(a) $d_1 = 350\mu\mathrm{m}$ (b) $d_1 = 600\mu\mathrm{m}$

(c) $d_1 = 1100\mu\mathrm{m}$ (d) $d_1 = 2600\mu\mathrm{m}$

Figure 4.31: *Intensity cross-section profiles of pinholes oriented along the horizontal axis and along the vertical axis of the central part of the white-light continuum beam. Position #1 indicated in Fig.(4.29).*

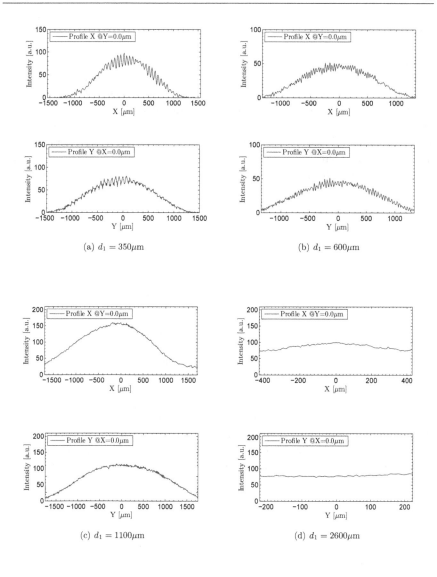

Figure 4.32: *Intensity cross-section profiles of pinholes oriented along the horizontal axis and along the vertical axis of the outer part (conical emission) of the white-light continuum beam. Position #2 indicated in Fig.(4.29).*

The plot of the visibility V versus the pinhole separation d, for the central part of the white-light beam is displayed in Fig.(4.33). At the separation of $d_1 = 350\mu m$ the beam exhibits only a visibility $V_x = 28.78\%$ on the horizontal axis, compared to the $V_{x,laser} = 83\%$ visibility of the pump-laser at this separation. On the vertical beam axis the visibility reaches only the value of $V_y = 23.17\%$. Remember that, for this pinhole separation, the pump-laser was completely coherent. This is a significant reduction in visibility and, thus ,degree of spatial coherence. The horizontal coherence seems to decrease faster than the vertical. This coincides with the observation of fringe visibility for the pump-laser. With increasing hole separation, the visibility of the fringes further drops down to about $V_{x,y} \approx 6\%$ for both beam axis. At the maximum pinhole separation, no fringes were seen, hence the spatial coherence at the center is zero, in both orientation.

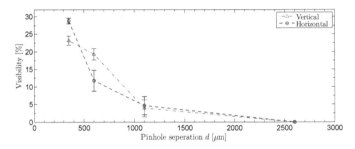

Figure 4.33: *Plot of the fringe visibility (%) as a function of pinhole separation, for horizontal beam axis (pinhole orientation)(dash-dotted) and vertical axis (dashed) of the white-light central part.*

The analysis of the fringe visibility of the beams outer part, where the conical emission is located, is shown in Fig.(4.34). At the smallest pinhole separation the visibility is already reduced about another 10%, compared to the central part of the beam. The horizontal and vertical axis show the same trend in visibility reduction. The spatial coherence is thus, significantly reduced compared to the center of the white-light beam, similarly reported in [GSE86], [Gol90].

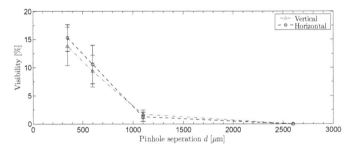

Figure 4.34: *Plot of the fringe visibility (%) as a function of pinhole separation, for horizontal beam axis (pinhole orientation)(dash-dotted) and vertical axis (dashed) of the white-light outer part (conical emission).*

Note that, during the measurement of the interference patterns with the CCD camera, the fringe contrast showed showed large fluctuations. Hence, each data set shows only a finite moment of the beams spatial coherence. Something similar has already been mentioned in section 4.4, where the positions of the filaments inside the crystal did not seem to be stationary. Therefore, it seems as if each pump-laser pulse breaks up in its own set of filaments, which then are located at different transverse positions. The total observed white-light beam, is thus, a superposition of many sources with large distribution of phases. This corresponds to the observation in [BIC96].

With the spatial coherence measurements at maximum input pulse energy of $E_0 \approx 270\mu J$, one can perform the following consideration. If the beam propagates with a finite diameter of $d = 25\mu m$ (as described in the introduction in chapter 2) inside the crystal and the diameter of the white-light spot is $d_{wl} \approx 2$cm of on the pinhole aperture, then two points $P_1(\vec{r}_1)$ & $P_2(\vec{r}_r)$ of the beam inside the crystal, separated by $\vec{r}_1 - \vec{r}_2 \approx 440$nm in the transverse plane, would exhibit a correlation of about 29%. At a separation of $\vec{r}_1 - \vec{r}_2 \approx 3.25\mu m$, no correlation would be found between the two points.

4.7 Conclusion

The intention behind this chapter was to find a suitable white-light continuum source in terms of spectral broadening, energy conversion efficiency and low spatial coherence for the generation of laser induced periodic surface structures on silicon and metal samples.

It could be shown that the white-light continuum with the highest light intensity is obtained for the maximum available input pulse energy of $E_0 \approx 270.5\mu J$. At this point the $\lambda_{laser} = 800$nm wavelength and $\Delta\lambda_{laser} = 15$nm spectral bandwidth is broadened to about $\Delta\lambda_{wl} = 650$nm, in the range of 350nm$\leq \lambda_{wl} \leq 1000$nm.

For the three filtered spectra, slope efficiencies of 12.8% (@800nm), 13.7% (@785nm) and 5.9% (@700nm) are obtained. At lower pump energies, the spectra were cut-off at longer wavelengths. Furthermore, it could be found that the white-light beam preserves the linear polarization of the pump-laser, however becomes slightly depolarized in the center at higher input energies. A long term stability measurement over a period of 20min. showed a 2.73% (1σ bounds) pulse to pulse stability.

The investigation of the spatial coherence in chapter 4.6 shows, that at maximum pump energy, the coherence of the white-light is significantly reduced.

A desired flat spectrum was only found around the central part of the white-light beam, and hence, cannot be selected individually. Therefore, an ablation spot will experience a steep spectrum (larger slope in narrow spectral intervals) in the center, which becomes flatter (smaller slope) towards the edge of the spot.

At the conclusion, the spectrum obtained at maximum pump energies, fits most of the desired needs for the LIPSS experiments in the following chapter.

Chapter 5

Self-Organized Pattern Formation with Ultrafast White-Light

A broad-band white-light (WL) spectrum (filtered, roughly ranging from $\approx$ 400nm$\leq \lambda_{wl} \leq$ 800nm), with significantly reduced spatial coherence, can only be obtained at the highest available input pump energy. For this reason, it was chosen to irradiate the samples, in the corresponding set of experiments, with a constant WL pulse energy and vary the irradiation dose $(I_D = f(N \cdot F))$, through the number of applied pulses N.

The main investigation will concentrate on the formation of structures on various materials upon WL irradiation. However, dielectric targets are not dealt with in this work, because of their high ablation threshold $(F_{die} \geq 10\mathrm{J/cm^2})$ and obvious limitations in the WL pulse energy and focusing of the WL (limited WL fluence). Therefore, materials with low ablation threshold such as silicon, copper, brass and stainless steel, $F_{Si} \approx 0.2\mathrm{J/cm^2}$ [JGL$^+$02], $F_{Cu} \approx 0.35\mathrm{J/cm^2}$, $F_{Brass} \approx 0.1\mathrm{J/cm^2}$ and $F_{Steel} \approx 0.16\mathrm{J/cm^2}$ [NSo10], [MPo00], correspondingly, were selected as targets.

Multi-pulse irradiation opens the possibility to the investigate the positive feedback effect of the ripple period lengths on metal samples, which is predicted to exhibit an exponential decreasing behavior (Eq.1.4) upon increasing irradiation dose.

Moreover, the analysis will focus on the observed characteristics and features of the produced LIPSS to identify similarities of LIPSS attributed to a self-organization process, from femtosecond-laser induced instability [Cos06], [Rei10], [Var13b].

The effect of the WL polarization on the direction of LIPSS is investigated, by placing a $\lambda/2$ retarder between lenses of the telescope arrangement (Fig.3.2) and attenuating its rotation angle, which changes the polarization direction of the WL.

Lastly, the impact of irradiation in air-atmosphere on the target surface element composition (oxidation, carbonation) of the silicon samples is demonstrated and discussed from EDX measurements.

To create LIPSS on the surface of the presented materials, the fluence of the incident light has to be at least in the magnitude of their ablation threshold F_{abl}. In order to achieve a sufficient high fluence the WL needs to be focused on a very small area. However, the large divergence angle of makes a focusing with a single lens difficult. For this reason the WL is focused with a telescope arrangement as shown in the setup in Fig.(3.2). On the downside, the usage of two spherical glass lenses results in a distortion of the focus (discussed in section 5.1) due to lens aberrations. Furthermore the WL pulse energy drops by another 16% of its peak, caused by the front and rear reflection of the light on each lens.

5.1 White-Light Beam Propagation

The focused white-light beam propagates in a very unique way, which is caused by the properties of the WL, like the beam filamentation and spatial dependance of the wavelength distribution. Hence, in this section the beam profile across different positions of the geometrical focus is monitored, in order to find a suitable position for subsequent target irradiation.

This could be done by taking CCD images of the transverse beam profile along its propagation axis (z-direction). The WL was dimmed with neutral density filters. Due to the short focal lengths of the lenses, the position of the focus region lies partially very close ($f_{wl} \leq 1$cm), to the second lens (f_2). Furthermore, the CCD-chip of the camera is mounted about 1.07cm inside the casing. This makes a direct measurement across the focus of the WL beam impossible. To overcome the difficulties, a *TV*-lens from *Electrophysics*, mounted directly onto the camera, was used to refocus the WL beam onto the CCD-chip at a distance of about 4cm from the second lens.

However, in order to get an information about the spot-size in [μm], the pixel size is calibrated for the adjusted zoom of the *TV*-lens by placing a 5mm×5mm grid, as well as a 5mm-scale in the focus of the lens. To reduce the error that accompanies this procedure, a mean value from the pixel values obtained from the grid and mm-scale was used. The gauge grid and mm-scale are shown in Fig.(5.1). The numbers on

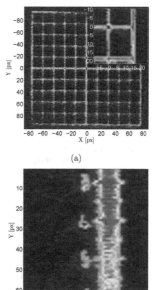

(a)

(b)

Figure 5.1: *CCD sensor pixel value gauge with (a) 5mm×5mm grid and (b) 5mm-scale. Numbers on scale do not correspond with mm-values.*

the mm-scale are not the mm-tick marks, here the distance between two major ticks is 0.5mm. The pixel unit size of the CCD sensor is specified as 8.6μm horizontally and 8.3μm vertically [Son14]. Now, with the *TV*-lens and the grid, the gauge values are, 0.5mm=15px horizontally and 0.5mm=17.5px vertically. The corresponding values from the mm-scale are 0.5mm=16px horizontally and 0.5mm=19px vertically. With that, the mean pixel sizes are $\bar{x}_{px} = 32.2\mu$m and $\bar{y}_{px} = 27.4\mu$m. The scales of the CCD images of the WL transverse beam profile in chapter 4.4 were obtained in the same manner.

Figure (5.2) shows the width of the WL beam, across the focus for the transverse x and y-directions. The beam width is displayed at the full width half maximum value. It is to notice that since the WL beam was refocused the z-axis is not comparable with the actual values of the geometrical focus. One can see already an asymmetric propagation of the focused beam. While approaching the waist, the width in horizontal and vertical transverse direction, decreases as expected. At the waist, the beam exhibits a size of $w_{0,x} = 104,4\mu$m in horizontal and $w_{0,y} = 103,3\mu$m in vertical direction at the FHWM. Yet, behind the waist in positive z-direction, the width becomes more irregular at about $z = +2mm$.

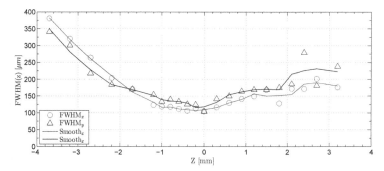

Figure 5.2: *Width of the white-light beam at the FWHM value in horizontal and vertical direction, across the recreated focus. z-axis is not comparable with the actual values of the geometrical focus.*

From this point the beam increases much slower in size. The smoothed data-sets show a general behavior of the beam's propagation. In figure (5.3), three transverse intensity profiles from different positions across the focus in z-direction are displayed.

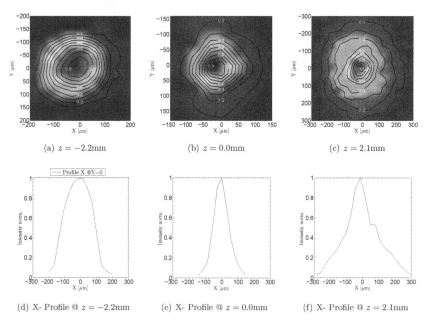

(a) $z = -2.2$mm   (b) $z = 0.0$mm   (c) $z = 2.1$mm

(d) X- Profile @ $z = -2.2$mm   (e) X- Profile @ $z = 0.0$mm   (f) X- Profile @ $z = 2.1$mm

Figure 5.3: *(a)-(c) Transverse intensity profiles of the white-light beam at 3 positions across the focused (z-scan). Positions correspond to the z-axis positions in Fig.(5.2). (d)-(f) Corresponding x-axis intensity profiles taken at $Y = 0\mu m$.*

The solid lines in the plots (5.3) are lines of equal intensity (i.e. isohypses) with intensity values equally increasing from $0.1I_{max}$ to $0.9I_{max}$.

One can see the isohypses in Fig.(5.3(a)) in almost equal distances. As the beam waist at $z = 0.0$mm is approached the intensity lines above $I > 0.2I_{max}$, still show the same behavior. However, at about the same distance from the waist in positive z-direction, isohypses of high intensity $(I > 0.5I_{max})$ are located closer around a peak near the center. Below the FWHM of $I < 0.5I_{max}$, the spread of the beam is larger and the isohypses are not in equal distance anymore and exhibit a more irregular shape, compared to the position before the waist. The corresponding x-axis intensity profiles, displayed in Fig.5.3(d)-(f), clearly show a significant change of the beam profile. The same behavior is visible in the ablation spots.

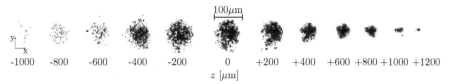

100μm

| y | x |

-1000   -800   -600   -400   -200   0   +200   +400   +600 +800 +1000 +1200

z [μm]

Figure 5.4: *Ablation traces of the focused white-light beam profile in longitudinal direction. The upper scale is only for the spot sizes not for the distance in z-direction. z-scale is not comparable to z-axis in Fig.(5.2)*

Figure (5.4) shows traces of the ablation spots produced on silicon after irradiation with $N = 1000$ pulses at a WL pulse energy of $E_d = 26.9\mu$J. The spots were taken $\Delta z = 200\mu$m apart, across the geometrical focus (z-scan). A definite asymmetric progression of the ablation is visible. The ablation spots consist of many individual hot-spots resulting from the beam's break up in many individual filaments. The first spot at $z = -1000\mu$m exhibits already a vertical spot diameter of about $d_y \approx 90\mu$m, but shows only very few ablated hot-spots. Here, only an effective area near the intensity maximum of the transverse beam intensity profile, is responsible for ablation. The effectivity of ablation starts to increase significantly at about $z = -600\mu$m. At the focus waist ($z = 0\mu$m) the spot diameters increased to about $d_y \approx 120\mu$m, $d_x \approx 95\mu$m. The corresponding area equals an effective area at $0.2 - 0.3I_{max}$ of the focused beam profile in Fig.(5.3(b)). In positive z-direction the spot diameter decreases much faster. At $z = +1000\mu$m the spot is $d_y \approx 30\mu$m, $d_x \approx 30\mu$m in diameter, and only the top of the central peak $(0.85 - 0.9I_{max})$ of the profile in Fig.(5.3(c)) is effective for the ablation.

A possible explanation for this beam propagation behavior could be the spatial dependence of the wavelengths inside the WL beam in connection with spherical and chromatic lens aberration. Due to the chromatic aberration, the shorter wavelengths exhibit a shorter focal length than the longer wavelengths. This results in the spread, and wavelengths dependence, of the focus. If one recalls the spatial properties in chapter 4.4, longer wavelengths are more intense near the center of the beam and shorter wavelengths are preferably located in the outer beam part.

Furthermore, the spread and dependence of the focus on the wavelengths, is additionally enhanced by the spherical aberration, caused by the spherical lenses in the telescope arrangement. Light rays located farther form the optical axis exhibit a shorter focal length, than rays near the optical axis.

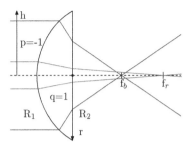

Figure 5.5: *Schematic of the +10mm spherical lens and light-ray propagation (from left to right). blue rays correspond to shorter wavelengths and red rays to longer wavelengths. R_1 front-side curvature radius, R_2 back-side curvature radius. r is the lateral lens radius. f_b focal point blue, f_r focal point red. q,p are "performing factors" of this lens type and orientation.*

An estimation of these aberrations for the 2nd lens in the telescope array (+10mm lens) show, that the measured behavior of the focused white-light beam may indeed be caused by chromatic and spherical lens aberration. The +10mm lens is fully illuminated during the experiments.

The spherical aberration, caused by the distance of the light rays h from the optical axis (dashed line in Fig.5.5), can be described by equation (5.1) [Dem13]. Where $p = (b-g)/(b+g) = -1$, called *shape factor* (b,g are the image and object distance, respectively), equals one for an object distance of $g \to \infty$. The *position factor* $q = (R_1 + R_2)/(R_2 - R_1) = 1$, equals one for a convex-plano lens ($R_2 = -\infty$, Fig.5.5). The factor n is the refractive index of the lens.

$$\Delta_s = \frac{h^2}{8f(h \approx 0)^3 n(n-1)^2} \left[n^3 + (3n + 2)(n - 1)^2 p^2 + 4(n^2 - 1)pq + (n + 2)q^2 \right]. \quad (5.1)$$

The deviation $\Delta_s = \frac{1}{f(h)} - \frac{1}{f(h \approx 0)}$, equals the difference between the reciprocal focal lengths for rays near the optical axis $f(h \approx 0)$, and for off-axis rays $f(h)$ at the distance h. The chromatic aberration is included via the "lens-makers equation" given by $f(h \approx 0) = f_0 = R_1/(n-1)$ (simplified for $R_2 = -\infty$) [Dem13].

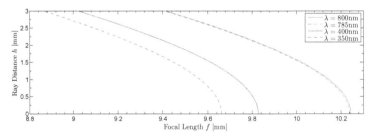

Figure 5.6: *Focal point curves for 4 different wavelengths (x-axis) in dependence on the light-ray distance h (y-axis), calculated with Eq.(5.1).*

Figure (5.6) shows the calculated focal point curves $f(h) = f_0/(f_0\Delta_s + 1)$ for 4 different wavelengths λ in dependence of the light-ray distance h. The lens specifications are $R_1 = 5.464mm$, $R_2 = -\infty mm$, $r = 3mm$ with an N-BaK2 material [Qio14]. The refractive indices of the 4 wavelengths in N-BaK2 are $n(\lambda = 350nm)=1.56557$, $n(\lambda = 400nm)=1.55595$, $n(\lambda = 785nm)=1.53369$, $n(\lambda = 800nm)=1.53337$ obtained from [Sch14].

As described, longer wavelengths are located in the beam center, while- shorter wavelengths are more intense in the beams outer part.

The focal point difference $\Delta f = f_r - f_b$ of an $\lambda = 800$nm and a $\lambda = 400$nm light-wave is already $\Delta f(\lambda_{800} - \lambda_{400}) = 1$mm, for $h = 2.5$mm ray distance of the $\lambda = 400$nm wave. For $\lambda = 350$nm it is $\Delta f(\lambda_{800} - \lambda_{350}) = 1.15$mm at $h = 2.5$mm, and $\Delta f(\lambda_{800} - \lambda_{350}) = 1.38$mm at the maximum $h = 3.0$mm.

These values in focal point difference, correspond to the z-scale magnitude in Fig.(5.4) and show that the focus spread and monitored irregular propagation behavior of the beam, may very well be the result of these aberrations. Furthermore, the aberrations take already place at the first lens (+20mm) in the array and enhance via error propagation in the 2nd lens (+10mm). This leads to the conclusion, that the ablation spots in front of the focus are created by a stronger blue-sided spectrum, while the spots behind the focus experience a spectrum with enhanced red-side. Correspondingly, the profile peak in Fig.(5.3(f)) is caused by longer wavelengths due to a longer focal distance. For this reason the irradiation position of most of the samples will be at the focus waist at around $z = 0.0\mu$m.

At this point it is to say that the spot shapes were very susceptible to the alignment of the telescope. A small misalignment of the two lenses, for instance in lateral direction, resulted already to a clubbed spot shape caused by *coma* aberration. The analysis of the focused WL beam from above represents the final alignments with almost circular spot shapes. However, during the experimental process the alignments was changed a couple of times and the first ablation spots recorded, show much more irregular shapes.

5.2 Image Processing and Period Length Determination

For the determination of the spatial period lengths of the ripples from the SEM- and AFM-images, a basic image processing technique is used. This involves fast Fourier transformation (FFT) in 2 dimension. The Fourier Transformation of an image $f(x, y)$ with dimension $M \times N$ is given by:

$$F(k_x, k_y) = \frac{1}{MN} \sum_{x=0}^{M-1} \sum_{y=0}^{N-1} f(x, y) \cdot e^{-i\left(\frac{k_x}{M}x + \frac{k_y}{N}y\right)} , \qquad (5.2)$$

with $F(k_x, k_y)$ being a complex function. The variables x and y are coordinates in space, whereas k_x and k_y are the coordinates in the reciprocal space also known as k-space. What this transformation basically does, is approximate the image $f(x, y)$ through sine and cosine wave-functions with different wavelengths and orientations of the waves. It then gives out an information which of the probed wavelengths are present in the image and how they are oriented.

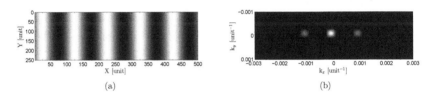

(a)   (b)

Figure 5.7: *(a) Sine-Wave pattern with defined wavelengths of λ=100, (b) Corresponding FFT image.*

In Fig.(5.7(a)) a periodic pattern is displayed, in this case a sine wave with defined wavelength of λ=100. The wavefronts of the pattern are oriented vertically in space. Figure (5.7(b)) shows the magnitude of the transformation $|F(k_x, k_y)|$. In k-space, wavelengths are related through $k = 1/\lambda$, thus longer wavelengths have lower k values. The central point, therefore, represents the approximation of the image in Fig.(5.7(a)) with large / infinite wavelengths, which is often associated with the background of an image.

The two opposite points, located left an right from the center, have in general the same information. They represent the only wavelength that exists in the image. At a value $k = 0.01$, where $k^{-1} = \lambda = 100$, which is indeed the spatial period length of the pattern. They are two points, because a sine and cosine wave can be described in positive and negative direction. This helps in particular to figure out the orientation of the pattern. Two points (regions) in the k-space, at equal distance and opposite direction from the center always represent a periodic wave pattern, with wavefronts oriented perpendicular to the secant line of these points.

In the general case of an image, very many wavelengths are approximated during the transformation. Thus, the k-space image consist of a large amount of points. In this case the global peaks in opposite direction represent the average wavelength of the dominant patterns.

Through the Fourier transformation of an image with periodic patters one can exactly determine the periodicity as well as the orientation of the pattern in space.

5.3 Structure Formation on Various Materials

The targets were conventionally cleansed with methanol and lens cleaning paper before irradiation. Afterwards some targets were post-processed in a 30°C heated ultrasonic bath, in order to remove of all redeposited material from the ablation in air. All ablation spots were produced under normal incidence ($\theta = 0°$).

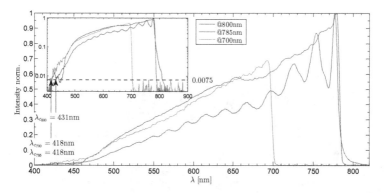

Figure 5.8: *Spectrum of the transmitted white-light continuum of all three short pass filter in the geometrical focus of the telescope arrangement. Inset shows log-scale plot.*

The filtered spectra in the focus of the telescope array are presented in Fig.(5.8). They were taken by placing a piece of paper in the geometrical focus (at about $z \approx$ 0.0mm, Fig.5.2) and recording the scattered light with the spectrometer.

All three spectra are falling off steeply from the max-

@800nm	431nm$\leq \lambda_{wl} \leq$810nm
@785nm	418nm$\leq \lambda_{wl} \leq$790nm
@700nm	418nm$\leq \lambda_{wl} \leq$703nm

Table 5.1: *WL Spectrum ranges for all 3 short pass filters.*

imum intensity near the cut-off. Here, a flatter spectrum (smaller slope in narrow spectral intervals) located in the beam's outer part is here overlaid by the much more intensive and steeper spectrum (larger slope) from the beam center. The ranges of the transmitted white-light continua, with relative intensities above $I/I_{max} \geq 0.75\%$ are shown in table (5.1).

The determination of the WL fluence is in the case of such a irregular beam propagation and spot shape, as shown above (Fig.5.2), more complicated than in a general case. Ablation spots taken in the focus are produced with different number of pulses under a constant WL pulse energy. Thereby, with the increasing number of pulses, the with LIPSS structured area increases as well and a larger effective area of the beam is responsible for ablation.

From the different structured areas, a mean value will be calculated. With the mean spot area, the fluence can be calculated through Eq.(3.1). However, the fluence is just an order of magnitude estimate in this work, since the intensity distribution of the transverse beam profile is not stationary and fluctuates, caused by the beam filamentation. In the case of very irregular shaped ablation spots, only the WL pulse energy will be given as reference.

5.3.1 Silicon

For the investigation of LIPSS on silicon various single-crystalline pieces from silicon-wafers were used. All of the silicon targets exhibited a (100) crystalline orientation, some were n-doped with metalloids like arsenic.

@800nm Filter

The first experiments on silicon were carried out with the @800nm cut-off filter. At this set of experiments, the focused white-light (WL) beam did not exhibit a circular transverse profile shape, caused from a misalignment of the telescope array as discussed in section 5.1, which can be noticed in the irregular spot shape consisting of several hot spots.

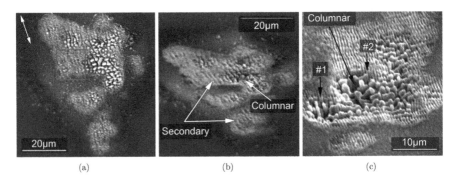

(a)   (b)   (c)

Figure 5.9: *LIPSS on (100) Silicon, produced with 100 WL pulses at a pulse energy of $E_d = 24.2\mu J$. (a) Overview of modified area (normal viewing angle), (b) 60° viewing angle, (c) 45° viewing angle (rotated 90° clockwise). The WL polarization is indicated by the double arrow. The red square in (a) indicates the enlarged region in Fig.(5.10(a))*

One of the first successful experiments of LIPSS formation upon irradiation with the WL continuum is presented in Fig.(5.9). Thereby, target was placed at about the waist (here defined as $z \approx 0.0$mm) of the focused beam and LIPSS were produced with 100 pulses at $E_d = 24.2\mu J$ per pulse.

The shape of the spot consists of several structured spots due to the local intensity variations. Large areas are structured with a primary ripple patten. In regions of high WL intensity, the ripples coalesce and coarser columnar structures are developed (indicated in Fig.5.9(b) and 5.9(c)).

The images in Fig.(5.9(b)) and (5.9(c)), show the spot under 60° & 45° viewing angles, correspondingly. The view parallel to the ripples in Fig.(5.9(b)) reveals a secondary pattern oriented perpendicular to the ripples, with a larger period length. Columnar structures in the center, are penetrating deeper in the material, than the ripples on the edge of the structured regions. Furthermore, the ripples near the craters extend all the way into the crater from which the columnar structures evolved (indicated #1,#2, Fig.5.9(c)).

Figure (5.10(a)) shows the enlarged square region from Fig.(5.9(a)).

The primary ripple pattern exhibits bifurcations (circle 5.10(a)) as well as truncations (square 5.10(a)) of lines. In between the ripple lines, holes are visible (indicated). The formation of such holes is a pre-phase in the evolution of a secondary pattern oriented perpendicular to the ripples.

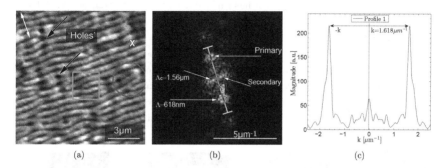

(a)   (b)   (c)

Figure 5.10: *(a) Enlarged square region from Fig.(5.9(a)). Square indicates region of ripple line truncation, circle ripple line bifurcation, the x marks the exact position in AFM an SEM image. (b) Magnitude image of FFT with indicated profile. (c) Cross section profile of FFT.*

The period lengths were determined via FFT, of the squared region in Fig.(5.9(a)). A linear order across the indicated profile is clearly visible in the magnitude FFT spectrum of Fig.(5.10(a)) shown in Fig.(5.10(b)). The extracted profile (Fig.5.10(c)), exhibits two main peaks, their distance from the center $k = 1.618\mu m^{-1}$, reveals the major spatial period length in Fig.(5.10(a)). By taking reciprocal value one obtains $\Lambda_{pri} = k^{-1} \approx 618nm^1$, which represents the average ripples period contained in Fig.(5.10(a)). The secondary pattern, oriented perpendicular to the primary, has a mean period length of $\Lambda_{sec} = 1.56\mu m$.

An investigation of the surface height profile is presented in Fig.(5.11). The AFM-scanned section is identical to the area of the SEM image in Fig.(5.10(a)).

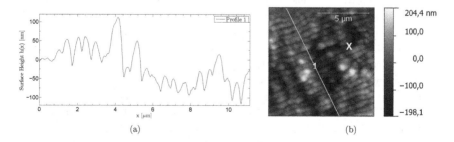

(a)   (b)

Figure 5.11: *(a) Extracted surface height profile, indicated in (b). (b) 10x10μm AFM image of square area from Fig.(5.9(a)). Position x = 0.0μm corresponds to the start of the profile at the upper edge in (b). The x marks the exact position in AFM an SEM image in Fig.(5.10(a)).*

---

[1]Throughout this work, the Fourier transformation between the k-space and the image is defined without 2π, hence $\Lambda = k^{-1}$.

The scaled AFM-image reveals a modulation of the surface height ranging from about -200nm up to +200nm, with respect to the mean surface level at 0.0nm in the upper right corner. The extracted profile (Fig.5.11(a)) shows the surface modulation perpendicular to the ripples and across the scanned region. The ripple height is about 50-100nm. Dark shaded areas in Fig.(5.11(b)) are valleys, which could not be identified as such in the SEM image (Fig.5.10(a)). In the lower left corner one can see the valleys being part of the coarser secondary pattern, oriented perpendicular to the ripples.

As comparison to the ablation spot in the focus and demonstration of the irregular beam propagation and spot shapes, an ablation spot at the distance of about +1000μm beyond the position of the spot from Fig.(5.9(a)) is presented in Fig.(5.12(a)). The irradiation conditions were the same as in the previous case. It is interesting to see how a deep crater formed. Ripples are only present on the edge of the crater, where the WL-intensity was lower, compared to the center.

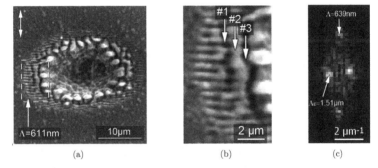

Figure 5.12: *(a) Ablation spot at about +1000μm beyond the spot position (Fig.5.9(a)) produced with 100 $E_d = 24.2\mu J$ pulses. (b) Close up of dashed area in (a). (c) Magnitude image of FFT of (b).*

Towards the center, a coalescence of the primary ripples into a coarser secondary and perpendicular oriented pattern is visible (indicated by arrows, Fig.5.12(a)). As the local WL intensity increases holes form (#1), ripple crests coalesce into a line perpendicular to the ripples (#2) and partially the interconnections between the holes disappears (#3 Fig.(5.12(b)).

The FFT image of Fig.(5.12(b)), shows two opposite points in vertical direction, which represent the average ripples period length at $\Lambda_{pri} = 639$nm, whereas the two peaks in horizontal direction represent the coarser secondary pattern with $\Lambda_{sec} = 1.51\mu$m (Fig.5.12(c)). These values are about equivalent with the measured spatial periods at the spot at $z = 0.0\mu$m relative position in the focus (Fig.5.10). Both spots were produced under the exact same irradiation conditions (100 $E_d = 24.2\mu J$ pulses). If one considers that this spot may experienced a red-enhanced WL spectrum (section 5.1), then no relation between the spatial ripple periods and the light wavelength can be found.

The coalescence of ripples into a coarser pattern, shows a dependence of the spatial periods on the local WL intensity (higher intensity in the center, recall Fig.5.3(c)), similar to structures on silicon produced with a monochromatic laser beam of Gaussian intensity profile [Var13b].

@785nm Filter

In a 2nd set of experiments the @785nm cut-off filter is used to reject the pump-laser from the continuum. From this point on the alignment of the telescope array and ablation spot shape is as presented in section 5.1. The samples were placed at the waist of the focus at $z = 0.0\mu m$ (Fig.5.4) and subsequently irradiated with a WL pulse energy of $E_d = 29.1\mu J$. Four spots were produced with 100, 250, 500 and 1000 pulses.

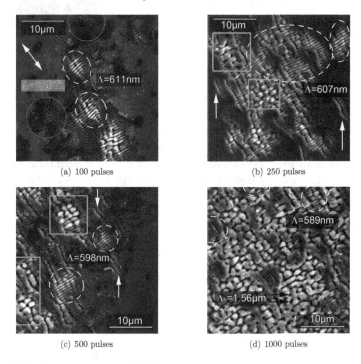

(a) 100 pulses   (b) 250 pulses

(c) 500 pulses   (d) 1000 pulses

Figure 5.13: *Dynamic investigation of LIPSS on multiple spots near their central part, produced with different number of pulses (F_{wl}=0.53J/cm², E_d = 29.1μJ) on silicon. The displayed spatial ripple periods Λ_{pri}, were obtained from the FFT (not displayed) of the dashed regions. (a) Solid circles show segments of early ripple formation (before a pronounced pattern developed). Double arrow indicates direction of WL polarization. Dashed circles show regions of a pronounced pattern. (b,c) Squares:, coalescence of primary ripples into coarser patterns. Arrows indicate isolated lines of holes. (d) Secondary periodic pattern with a hole periodicity of $\Lambda_{holes} \approx 1.56\mu m$.*

The average spot area, of those 4 spots, is $\bar{A}_{Si,spot} = 5310\mu m^2$, which corresponds to an effective area of the transverse beam profile in Fig.(5.3(b)), at about 60%-65% of the relative peak intensity (Fig.5.3(b)). Hence, the WL fluence has a value of $F_{wl} = 0.55J/cm^2$, which is $2.7 \times F_{Si}$ the ablation threshold of silicon. In figure (5.14), a typical ablation spot produced with 250 pulses is shown. Only a few areas of the modified area are covered with ripples. Furthermore, the structured areas have a preferential direction along the WL polarization.

The evolution of the primary ripple pattern upon multiple pulses is presented in Fig.(5.13). Upon 100 pulses, segments of ripples start to form (solid circles Fig.5.13(a)). The length of a ripple line is here below $< 3\mu m$. Thus patterns do not show ordering over several tens of microns, yet the ripples lines exhibit already bifurcations and truncations. An identical feature of short ripple lines is shown in the simulation results in Fig.(1.8) during the early appearance of the structures. The spatial period in the solid circles is $\Lambda_{pri} = 651nm$.

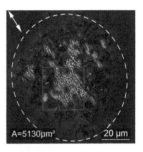

Figure 5.14: *Ablation spot produced with 250 $F_{wl} = 0.53J/cm^2$ pulses. Square indicates enlarged region in Fig.(5.13(b)). Dashed circle indicates affected area.*

Regions with a higher structure density were obviously exposed to higher WL intensity, resulting from the intensity variation in the beam filamentation. Here, a pronounced ripple pattern, with an average spatial period of $\Lambda_{pri} = 611nm$ formed (dashed circles Fig.5.13(a)).

Upon an increased number of pulses (250 & 500, Fig.5.13(b) and 5.13(c)), larger areas are structured with LIPSS. Unlike to the structured spots in Fig.(5.9(a)) and Fig.(5.12(a)), linear formation of the holes sometimes isolated from the ripple pattern, oriented parallel to the WL polarization could be found (indicated by arrows). In the regions indicated by the squares, the primary ripples coalesce into coarser structures, as a result of the multi-pulse feedback effect.

At 1000 pulses, most of the primary pattern disappeared and coarser structures with reduced linear order developed. These structures are separated by holes and ditches at a mean periodicity of $\Lambda_{holes} = 1.56\mu m$ (example indicated Fig.5.13(d)).

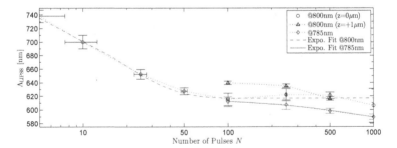

Figure 5.15: *Behavior of ripple period length in dependence of number of pulses, for the three discussed cases. (@800nm)Fit parameters $A_{800} = 161.9 \pm 12.1nm$, $y_0 = 616.8 \pm 3.0nm$, $t = 16.1 \pm 2.4$. (@785nm)Fit parameters $A_{785} = 39.1 \pm 8.0nm$, $y_0 = 577.8 \pm 10.5nm$, $t = 808.7 \pm 417.4$. The pulse error-bars results from the mechanical nature of the shutter, which is specified to open and close within 5ms. Roughly, the first ~2.5 and last ~2.5 pulses of each pulse train set may be cut-off.*

The behavior of the primary ripple period lengths observed on silicon is presented in Fig.(5.15). A clear decreasing tendency of the periods at an increased irradiation dose (constant fluence, number of pulses changed) is found. Here, the data of the (@800nm, Fig.5.9) and (@785nm, Fig.5.13) periods were approximated by an exponential fit, given by Eq.(5.3):

$$\Lambda_{\text{LIPSS}} \propto A \, \exp\left\{-\frac{N}{t}\right\} + y_0, \qquad (5.3)$$

with A, t being the fit parameters and N the number of applied pulses. The fit shows a good agreement with the data and a definite exponential decreasing tendency of the primary ripple periods, just as predicted by the *dynamic*-model in the linear regime (chapter 1.2.3) via,

$$\Lambda_{\text{LIPSS}} \propto \frac{1}{F} \, \exp\left\{-\frac{E_a}{2aF}\right\} . \qquad (5.4)$$

At an increased fluence, instantly smaller periods were measured, than compared to periods at the same number of pulses, yet slightly lower fluence. This further demonstrates the reciprocal (Eq.5.4) dependency of the ripple periods on the fluence of the WL pulses.

For the spot at $\approx +1000\mu$m from the focus (Fig.5.12), the ripple periods also decrease with increasing number of pulses, however, above $N > 500$ pulses the ripple pattern is fully washed-out.

With the primary ripple pattern being part of the linear regime, is the observation in good agreement with the *dynamic*-model of LIPSS formation. In general, $\Lambda_{\text{prim.}}$ have been found to vary in the range 600nm-750nm on silicon for multi-pulse irradiation. Coarser and secondary patterns exhibited spatial periods above $> 1\mu$m and, thus, above the range of the WL spectrum. Similar pattern periods are reported in [Var13b], [Cos06], which are explained in the framework of self-organized pattern formation (*dynamic*-model).

5.3.2 Stainless Steel

@785nm Filter

The stainless steel samples were mechanically polished by hand. Thus, they exhibit scratches of different size on the surface. Nevertheless, this could be of scientific interest, because it has been published [SSt82], [RVC08], that ripples are influenced by scratches and surface defects. The irradiation conditions are similar to the silicon section, with $E_d = 29.1\mu J$ WL pulse energy.

Since stainless steel has a lower ablation threshold ($F_{Steel} \approx 0.16\text{J/cm}^2$, [JGL$^+$02]) than silicon ($F_{Si} \approx 0.2\text{J/cm}^2$), a larger area of the beam profile is now effective on the surface. Seven spots were produced with a number of pulses from 25 up to 10.000. They exhibit a mean structured area of $\bar{A}_{Steel,spot} = 9210\mu m^2$. This corresponds to an effective beam area at about 30%-35% of the peak intensity (Fig.5.3(b)). At the central part of the spot, the beam filamentation does not play a significant role, since the variations from the normalized peak intensity are only about $\Delta I/I \approx 50\%$ (recall Fig.4.16(c)). The applied fluence corresponds to a value of $F_{wl} = 0.32\text{J/cm}^2$, which is $2\times F_{Steel}$ the ablation threshold of stainless steel.

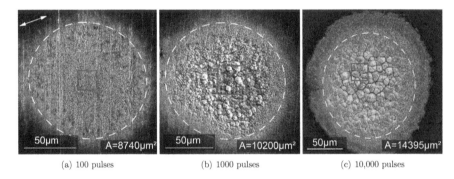

(a) 100 pulses   (b) 1000 pulses   (c) 10,000 pulses

Figure 5.16: *Ablation spots produced on stainless steel with $F_{wl}=0.32$ J/cm^2. Dashed circle indicates structured spot area used for fluence calculation. Squares indicate enlarged region in Fig.5.17(b), Fig.5.17(e), Fig.5.17(f).*

Typical structured spots are shown in Fig.(5.16). At 100 pulses most of the indicated dashed area is covered with the primary ripple pattern (edge exhibits structured spots, caused by the beam filamentation). At increased irradiation dose, coarser secondary structures, first in bubble-like shape (1000 pulses) and later (10,000 pulses) in the shape of large patches, are developed. The enlarged details of the spots central part is presented in Fig.(5.17 b,e,f). All displayed images in figure (5.17), were taken at about the same relative position, near the center, of the individual spot.

The effect of the surface scratches is demonstrated in Fig.(5.17(a) and 5.17(b))). At 50 pulses, ripples indeed start to form in the vicinity of the vertical scratches, preserving their orientation (indicated). The mean ripple period length in regions away from the scratches is $\Lambda_{pri} = 423\text{nm}$, at the scratches it is $\Lambda_{scratch} = 371\text{nm}$.

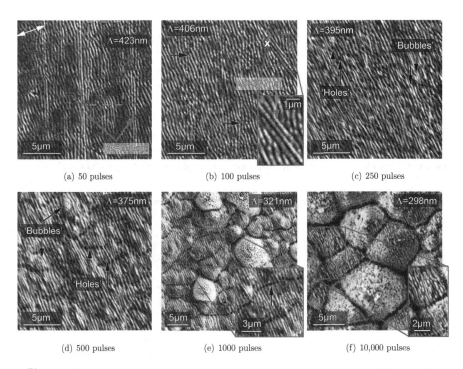

Figure 5.17: *Dynamic investigation of ripples periods on multiple spots, produced with different number of ($F_{wl} = 0.32\,J/cm^2$, $E_d = 29.1\mu J$) pulses on stainless steel. The displayed spatial ripple and scratch periods $\Lambda_{pri,scratch}$, were obtained from the FFT (not displayed). (a) Spatial period obtained from dashed region via FFT. (b) Black arrows show vertical surface scratches intersecting ripple lines, the x marks the exact position in AFM Fig.5.18 and SEM image. Magnified square inset shows alignment of ripples to the scratch (c,d) Secondary pattern effects, early formation of holes and bubbles. (e,f) Inset shows magnified squared region. Arrow indicates line of holes.*

At 100 pulses the center of the spot is evenly covered with a ripple pattern of $\Lambda_{pri} = 406$nm period length. In the upper right corner of Fig.(5.17(b)), one can see that the ripples are aligned along the large scratch, insensitive of the WL polarization. Minor scratches in horizontal direction intersect the ripple pattern (indicated, arrows).

The average ripple length (or "bifurcation length"), e.g. length of line structure before it bifurcates or truncates, is below $< 5\mu$m. Thereby, the ripples on the larger scratch exhibit a $\Lambda_{scratch} = 372$nm period. Similar to the case at 50 pulses, this shows that, the periods tend to be smaller on scratches, which may result from enhanced absorption at this site [SSt82].

Above $N > 250$ pulses holes start to appear, indicating the formation of a secondary pattern, similar to the case of silicon. This pattern is in general also oriented perpendicular to the ripples, yet does not exhibit such a strong linear order (Fig.5.17(d) and 5.17(e)). On stainless steel, at an increased irradiation dose, coarser structures manifest themselves as bubble-like objects, which are covered with ripples (Fig.5.17(d) and 5.17(e)).

At 1000 pulses, spatial periods in a $\Delta\lambda_{pri} \approx 100$nm range, coexist in the same region with an average value of $\bar\Lambda_{pri} = 321$nm (Fig.5.17(e)). The segments of the secondary hole pattern, are now interconnected (indicated arrow in inset Fig.5.17(e)).

After 10,000 pulses, a weak ripple pattern with $\Lambda_{pri} = 298$nm mean period is found on the patches. Thereby, the hole pattern developed into a groove pattern, while interconnections of ripple lines extend beyond the grooves (Fig.5.17(f)). The bifurcation length decreases significantly at the increased number of pulses and, thus, appears to be strongly correlated to the positive feedback effect [RVC08].

In figure (5.18) the investigation of the surface height profile of the spot with 100 pulses is displayed. The selected part of the AFM-image is identical to the SEM image in Fig.(5.17(b)).

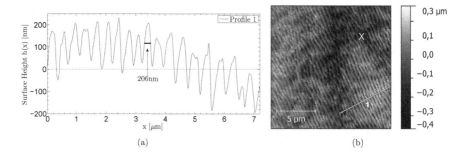

(a)   (b)

Figure 5.18: *Scanned AFM image from spot (Fig.5.17(b)) produced with 100 ($F_{wl} = 0.32J/cm^2$, $E_d = 29.1\mu J$) pulses. (a) Extracted surface height profile, indicated in (b). (b) 15x15μm AFM image comparable with SEM, the x marks the exact same position.*

Similar to ripples produced on silicon with 100 pulses, on steel the ripples grew about +300nm above the mean surface level, too. Deep valleys up to -400nm are only found in the vicinity of the surface scratches, visible here as dark shaded areas. In average, the ripple crest show a width of $\bar d_{crest} = 221$nm, measured at the FWHM. Some of them exhibit a split top.

@700nm Filter

Due to the low ablation threshold of stainless steel, it was possible to produce LIPSS with the @700nm cut-off filter, demonstrated in Fig.(5.19). 16% of the light transmitted by the @700nm filter, were reflected by the telescope arrangement. Hence, by pumping the Al_2O_3 crystal with pulses of $E_{p,laser} = 270.5\mu J$, a WL pulse energy of about $E_{p,wl} = 15.4\mu J$ could be obtained.

Above a number of 500 pulses a primary ripple pattern could be found, shown in Fig.(5.19(a)). The magnitude FFT inset shows distinct presence of a linear order. It can be seen that not one, but a range of period lengths is present (indicated). On the average period lengths found around $\bar\Lambda_{pri} = 295$nm. Here, the average bifurcation length is below $< 2.5\mu m$ and the pattern exhibits a low degree of regularity.

Upon 1000 pulses, the average ripple period length exhibits a value of $\Lambda_{pri} = 330$nm, which is slightly higher.

However, the ripples show a progressive reduction in their regularity (cf. inset, Fig.5.19(a)), visible on the inset in Fig.(5.19(b)). Because of the low WL pulse energy, a spot with a higher number of 60,000 pulses was produced as well. As shown in Fig.(5.19(c)), The surface appears strongly eroded. The center exhibits larger bubble-like structures. Furthermore, no extended ripple pattern can be identified (Fig.5.19(c)). On the edge of the same spot, a linear pattern parallel to the polarization of the WL was found (Fig.5.19(d)).

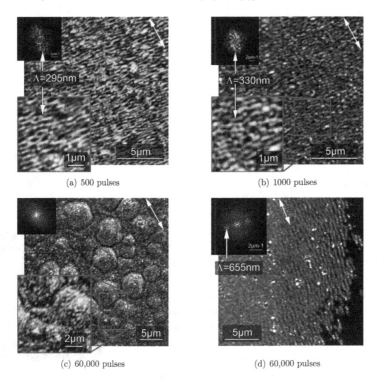

(a) 500 pulses   (b) 1000 pulses

(c) 60,000 pulses   (d) 60,000 pulses

Figure 5.19: *LIPSS produced on stainless steel with the @700nm cut-off filter and $E_{p,wl} = 15.4\mu J$ pulses. Double arrow indicates direction of WL polarization. (a,b) central region of a spot produced with 500 & 1000 pulses. The displayed spatial ripple periods Λ_{pri}, were obtained from the FFT (upper left corner) of the magnified region. (a) Inset shows magnified lower left image corner. (c,d) center, edge of the same spot, respectively, produced with 60,000 pulses. (c) Inset FFT of entire image.*

This pattern exhibits a period length of $\Lambda_{edge} = 655$nm and was only visible after cleansing the sample in an ultrasonic bath.

The tendency of the ripple period lengths, found on stainless steel, under the presented irradiation conditions is shown in Fig.(5.20). In general, a decreasing tendency of the periods, as observed on silicon (Fig.5.15), is visible. The fit shows, again, the exponential decay, which demonstrates a strong correlation to the theoretical prediction (Eq.1.4) for metal samples by the *dynamic*-model. In the case of LIPSS produced with the @700nm filter no significant statement about the behavior of ripple periods can be made. A reason for the significantly different appearance of the LIPSS produced with the @700nm filter could be the changed irradiation conditions, of a lower white-light pulse energy and different focusing dynamics of the white-light, since longer wavelengths located in the center of the WL beam are blocked out.

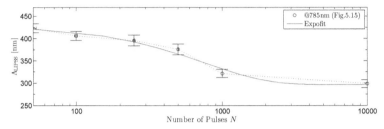

Figure 5.20: *Behavior of ripple period length on stainless steel, in dependence of number of pulses, produced with ($F_{wl} = 0.32 J/cm^2$, $E_d = 29.1\mu J$) pulses and the @785nm cut-off filter. Exponential Fit with Eq.(5.3), fit parameters are $A_{785} = 133.9 \pm 12.1 nm$, $y_0 = 296.2 \pm 9.7 nm$, $t = 752.0 \pm 167.7$.*

5.3.3 Copper

@785nm Filter

Copper, of all investigated targets has the highest ablation threshold of $F_{Cu} \approx 0.35 J/cm^2$. It was irradiated with a WL pulse energy of $E_d = 29.1\mu J$. Since E_d is already the highest possible pulse energy that could be achieved, a higher dose (i.e. larger amount of pulses), was applied to produce a pronounced ripple pattern on the copper surface. The WL stability analysis (chapter 4.5) showed, that a stable WL pulse energy with a 2.7% pulse to pulse deviation is guaranteed.

Ablation spots were produced with a number of pulses ranging from $N = 1000$ to $150,000$, while first indications of LIPSS could be found above 15,000 (Fig.5.22(a)). The spots exhibited a mean structured area of $\bar{A}_{Cu,spot} = 13717\mu m^2$, which corresponds to an area at about 25-30% of the beam's maximum intensity in the focus (Fig.5.3(b)). Hence, the WL fluence exhibited a value of $F_{wl} = 0.21 J cm^2$, which is about $0.6 \times F_{Cu}$ the threshold of copper. Typical spots produced of copper are shown in Fig.(5.21). The vertical lines, visible in the SEM images, are again scratches from the mechanical polishing.

After 15.000 pulses a modification of the surface is already visible, however, the existing pattern is still dominated by the scratches on the surface (Fig.5.21(a)).

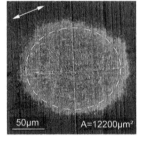

(a) 15,000 pulses

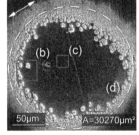

(b) 150,000 pulses

Figure 5.21: *Ablation spots on copper with $F_{wl} = 0.21 J/cm^2$. Squares enlarged in Fig.(5.22).*

By applying ten times more pulses, a large modified spot was observed, with irregular clustered structures on its edge. Here, a steady ripple pattern was found near the edge of the inner region of the spot (Fig.5.21(b)).

In figure (5.22(a)), the center of the spot produced with $N = 15,000$ is shown. The surface appears congealed from a previous molten-liquid state, recognizable through the round "droplet-like", edges. Here, periodic structures show an average period of $\Lambda_{pri} = 516$nm. Occasionally, larger holes with a diameter of $\bar{d}_{holes} \approx 600$nm can be found randomly scattered, over the modified area of the crater (Fig.5.22(a)).

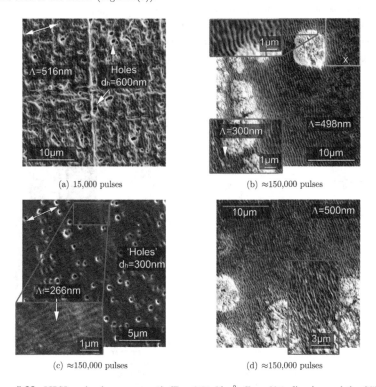

(a) 15,000 pulses   (b) ≈150,000 pulses

(c) ≈150,000 pulses   (d) ≈150,000 pulses

Figure 5.22: *LIPSS produced on copper with ($F_{wl}=0.21$ J/cm², $E_d = 29.1\mu J$) pulses and the @785nm cut-off filter. The displayed spatial ripple periods Λ_{pri}, were obtained from the FFT (not displayed). Double arrow indicates direction of WL polarization, and is the same in all 4 images. (a) Central part of spot produced with 15,000 pulses. Average hole diameter is $\bar{d}_{holes} \approx 600$nm. (b) Left edge of inner spot region, inset show magnified lower left corner. Green square indicates AFM (Fig.5.23) scanned area, while x marks the exact position in AFM and SEM image. (c) Center of the same spot. (f) Right edge of inner spot region of the same spot as in (b,c). Inset show magnified lower area.*

In [Var13b], [MCN⁺96], similar holes were reported on silicon before a pronounced ripple pattern evolved. It is suggested that such holes could arise from plasma formation in the defect states on the crystal lattice, followed by an expansion into the bulk, which results from a strong recoil pressure.

Figure (5.22(b)-(d)), shows SEM micrographs taken at different positions of the same ablation spot produced with $N \approx 150.000$. The primary ripple pattern on the edge of the inner crater region behaves in a peculiar way. An almost evenly transition of ripple periods near the center to the edge with $\Lambda_{pri} = 498$nm to $\Lambda_{pri} \approx 300$nm, correspondingly, is visible. The bifurcation length is about $\approx 10\mu$m, however, towards the edge, ripples exhibit a progressive reduction in regularity and the bifurcation length reduces to about $\approx 2\mu$m.

Similar to the case of silicon, it seems that the holes indicate a coarser secondary pattern oriented perpendicular to the ripple pattern with a period of $\Lambda_{sec} = 1.3\mu$m. In between the small holes ripple crests coalesce. This can be seen well in AFM image (indicated arrow in Fig.(5.23). The displayed AFM image equals the square in the upper right corner in Fig.(5.22(b)). In the SEM image, one can see how ripple lines are influenced by these structures and how they bend towards them (inset upper left corner in Fig.(5.22(b))).

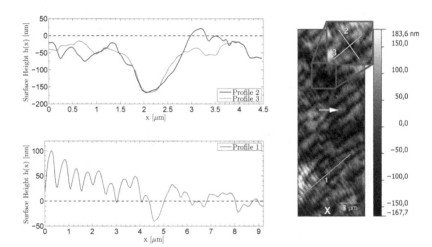

Figure 5.23: *Surface height profile on copper ($10^5\#$, $F_{wl} = 0.21 J/cm^2$). (r.) Scaled AFM image equals square in upper right corner in Fig (5.22(b)). Arrow indicates ripple coalescence. (l. upper) Cross-section height profiles 2 and 3 through larger hole. The profiles show the real depth and diameter, since the cantilever-tip is 15μm long with a 15° opening angle. (l. lower) Extracted surface height profile 1. The x marks the exact same position as in the SEM (Fig.5.22(b)) image.*

The surface height *profile 1* (Fig.5.23 lower plot) further reveals, that ripples in the vicinity of such objects are also located higher above the surface level than the surrounding structures. The inset (Fig.5.23 AFM image, right), demonstrates the cross-section height profiles through one of the larger holes mentioned above. It has been found that their depth is about $h \approx -100$nm.

The central part of the crater (Fig.5.22(c)), appears smooth and exhibits randomly scattered holes with a diameter of about $\bar{d}_{holes} \approx 300$nm.

Moreover, at the details a very fine ripple pattern oriented parallel to the polarization of the light with a period length of $\Lambda_{fine} = 266$nm. This is displayed as a magnified inset of the dashed square in Fig.(5.22(c)).

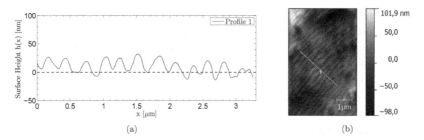

(a)　　　　　　　　　　　　　(b)

Figure 5.24: *Surface height investigation of fine ripple pattern found in the center of the crater. (a) Extracted surface height profile 1, (b) scaled AFM image.*

An investigation of the surface height in Fig.(5.24(a)), shows that the ripple height is only about $h = 25$nm and thereby much smaller the the regular ripple pattern found on the inner edge of the spot.

Figure (5.22(d)) shows again the edge of the inner crater region at a different position. Here, it can be seen how the spatial period of the ripples smoothly decreases perpendicular to the edge of the crater.

@700nm Filter

Due to a smaller pulse energy of $E_d = 15.4\mu J$ around $N \approx 300,000$ pulses were applied to produce LIPSS with the shorter wavelength spectrum. The three SEM images in Fig.(5.25) represent three different locations of the same spot.

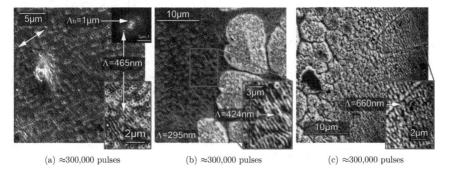

(a) ≈300,000 pulses　　　(b) ≈300,000 pulses　　　(c) ≈300,000 pulses

Figure 5.25: *LIPSS produced on copper with 300,000 $E_d = 15.4\mu J$ pulses and the @700nm filter. The displayed spatial ripple periods Λ_{pri}, were obtained from the FFT (not displayed in (b,c)). (a) Central part of spot. Inset shows magnified lower left region, with ripple pattern (indicated in FFT as well) and hole pattern perpendicular to ripples and WL polarization. (b) Edge of inner spot region. (c) Edge of the same spot.*

In the center of the spot shown in Fig.(5.25(a)), a ripple pattern with a spatial period Λ_{pri} = 465nm, oriented perpendicular to the WL polarization could be found (indicated, arrows, magnitude FFT image).

The holes in the center of the spot, form a distinct pattern with Λ_{holes} = 1μm average period (5.25(a) indicated, arrows, magnitude FFT image).

Similar to the previous spot (Fig.5.22(b)), the edge of the inner spot region exhibits a primary ripple pattern with a period length of Λ_{pri} = 424nm. Between the coarse round objects, the spatial period decreased down to Λ_{ripple} = 295nm (indicated, Fig.5.25(b)).

Outside the cater, a very irregular linear pattern with a period of Λ = 660nm, oriented perpendicular to the WL polarization was found. Like in the case of stainless steel, this pattern was revealed after ultrasonic cleaning the sample (Fig.5.25(c)).

The tendency of decreasing periods with increasing number of applied pulses at a constant pulse energy, as observed on silicon and stainless steel, can be found on copper as well. Additionally, the first pronounced ripple pattern produced with a WL spectrum in the range of λ_{wl} = 418 − 703nm (Fig.5.8) could be identified. In the spectrum (Fig.5.8), wavelengths around λ_{wl} = 690nm, exhibit the highest intensity. Assuming these high intensity wavelengths are responsible for the ripple formation, then the measured Λ_{pri} = 420nm period length is still 270nm smaller. Around λ_{wl} = 420nm, the continuum only has a relative intensity of $\sim$ 1%, hence a relation between the wavelength of the incident light and the ripple period, predicted by the *static*-model (chapter 1.2.2), cannot be identified.

5.3.4 Brass

@785nm Filter

The last of the investigated materials is a brass sample. Brass is an alloy of the transition metals copper (Cu) and zinc (Zn). The material properties of brass, for example melting point and hardness, can be influenced by the percentage of Cu and Zn composing the brass ($CuZn_X$), where X holds the zinc percentage. The sample is industrial brass ($CuZn_{>60}$), with a high zinc content of over ($X > 60\%$).

With a WL pulse energy of E_d = 29.1μJ 3 Spots were produced with a number of pulses ranging from N = 500 − 15,000. The modified surface area thereby exhibited an average value of $\bar{A}_{Brass,spot}$ = 9770μm^2, which corresponds to an effective beam area at about 30% the peak intensity. Thus, the applied fluence is F_{wl} = 0.29J/cm^2, which is 2.9 × F_{Brass} the ablation threshold of brass.

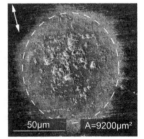

(a) 500 pulses

Figure 5.26:
Ablation spot on Brass produced with F_{wl} = 0.29J/cm^2.

Typical ablation spots observed on brass are presented in Fig.(5.26). For a low number of pulses the effect of the beam filamentation and spatially inhomogeneous structuring is visible (Fig.5.26(a)), similar to on silicon and copper.

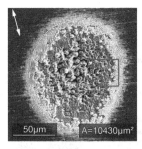

(a) 15,000 pulses

Figure 5.27:
Ablation spot on Brass produced with $F_{wl} = 0.29J/cm^2$.

At an increased number of pulses, the surface is heavily modified, exhibiting large ditches and holes (visible in Fig.5.27(a)).

After 1000 pulses, a weak primary ripples pattern is developed with a spatial period length of $\Lambda_{pri} = 506$nm in the central area of the spot (enlarged inset, Fig.5.26(a)). Randomly distributed holes can be identified, similar to the case of the elementary copper sample.

At an increased dose ($N = 10,000$, Fig.5.28(b)), ripples were only found at the edge of the inner spot region with a period of $\Lambda_{pri} = 500$nm. The holes are further evolved into irregular ditches (indicated). Additional, linear arrangements of holes oriented parallel to the WL polarization, similar to the ones on silicon, are observed (indicated arrows, Fig.5.28(a)).

After $N = 15,000$ the spatial period lengths of the ripple pattern decreased to a value of $\Lambda_{pri} = 472$nm. The large ditches as well as the holes, are a sign of dominating thermal effects during irradiation. Also on brass, a tendency of decreasing ripple periods, correlated to a multipulse feedback could be observed.

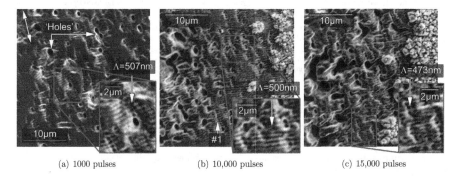

(a) 1000 pulses   (b) 10,000 pulses   (c) 15,000 pulses

Figure 5.28: *LIPSS produced on Brass ($CuZn_{>60}$) produced with ($F_{wl} = 0.29J/cm^2$, $E_d = 29.1\mu J$) pulses and the @800nm filter. The displayed spatial ripple periods Λ_{pri}, were obtained from the FFT (not displayed) of the magnified regions. Double arrow indicates direction of WL polarization. (a) Central region of spot with 1000 pulses. Randomly distributed holes indicated. (b) Central and edge of inner spot region. Isolated line of holes indicated. (c) Central and edge of inner spot region of an affected area produced with 15,000 pulses.*

@700nm Filter

With the @700nm cut-off filter, two spots with 30.000 and 120.000 pulses were produced at an energy of $E_{p,wl} = 15.4\mu J$. Upon a smaller number of applied pulses, no periodic structures could be observed. Figure (5.29(a)) shows a part of the center and the edge of the spot after 30.000 pulses.

A periodic ripple pattern with $\Lambda_{\mathrm{ripple}} = 520\mathrm{nm}$ period length could be observed. Ripples are more pronounced on the edge of the spot and are weaker close to the center. The irradiation dose ($30,000$ $E_{p,wl} = 15.4\mu$ pulses) was about the same as on the spot produced with $15,000$, $E_d = 29.1\mu\mathrm{J}$ pulses and the @785nm filter (Fig.5.28(c)). However, the patterns do not look alike. As previously discussed, through the lower spectrum cut-off, the focusing dynamics in the telescope array change as well, which further has an effect on the focused area and WL fluence. Hence the interaction is weaker and a weaker pattern evolves.

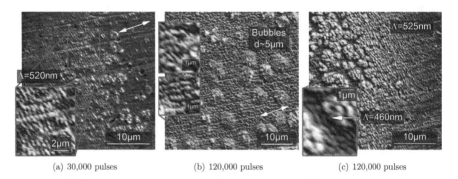

(a) 30,000 pulses   (b) 120,000 pulses   (c) 120,000 pulses

Figure 5.29: *LIPSS on Brass produced with $E_{p,wl} = 15.4\mu J$ pulses and with the @700nm cut-off filter. The displayed spatial ripple periods Λ_{pri}, were obtained from the FFT (not displayed) of the magnified regions. Double arrow indicates direction of laser polarization. (a) Edge of a spot produced with 30,000 pulses Inset shows magnified square region. (b) Central region of a spot produced with 120,000 pulses. Average bubble diameter is about $\bar{d}_{bubble} = 5\mu m$. Insets show magnified square regions with remains of a periodic ripple pattern. (c) Edge of the same spot.*

At 120.000 pulses, instead of deep ditches and holes, the center exhibits coarse "bubble-like" structures which are scattered randomly (Fig.5.29(b)). These "'bubbles'" have an average diameter of $\bar{d}_{bubble} = 5\mu m$. Similar bubbles were observed on dielectrics, attributed to pressure-induced compressive stress due to fast laser heating, followed by the relaxation of the surface [Var13b].

Additionally, the surface is covered with smaller bubbles-like structures. In a few places a periodic pattern can be identified (shown on the two insets in Fig.5.29(b)). It seems that an initial ripple pattern already coalesced into a secondary pattern.

On the edge of the same spot, the remnants of this pattern (inset in Fig.5.29(c)) are visible. The spatial period of the pattern is $\Lambda_{\mathrm{pri}} = 460\mathrm{nm}$ and can be found in between the coarser structures. The linear pattern at the outer edge shows a mean period of $\Lambda_{\mathrm{edge}} = 525\mathrm{nm}$.

5.4 Effect of White-Light Polarization

In the previous sections it has been shown that the orientation of the primary ripple pattern is perpendicular to the polarization axis of the white-light. In most cases the secondary pattern, which evolved directly from the ripples, exhibited an orientation parallel to the polarization of the light. The orientation of the primary ripple pattern can be explained in the framework of the *dynamic* (self-organization) model of LIPSS formation [Var13b].

In principle, it is stated that the polarization dependence is mainly caused by an anisotropic energy transfer of the excited electrons during irradiation, which results from an asymmetry in the ionized kinetic energy distribution. This anisotropic energy transfer leads to larger velocity of electrons in field direction than off-field. However, with higher kinetic energy the mean free path of the electron is significantly reduced [Rei10]. Thus, for electrons moving in field direction the collision probability with a phonon is much higher, than for electrons moving perpendicular to the field. As a consequence lines oriented perpendicular to the electric field of the incident light, form during the self-organization phase of the surface. This means that by changing the direction of polarization, one can control the orientation of the primary ripple pattern.

In chapter 4.3 it was discussed that the WL continuum preserves the linear polarization of the pump-laser beam. Hence, in the following experiment a $\lambda/2$ wave-plate was placed in between the two lenses of the telescope array, shown in Fig.(3.2) and the polarization direction of the WL was changed by attenuating the rotation angle of the wave-plate in 30° steps, from $\alpha = 0°$ to $\alpha = 90°$.

In figure (5.30) a continuous ripple pattern, from top to bottom, produced on silicon with N=700 (@785nm filter,$E_d = 29.1\mu J$, $F_{wl} = 0.55 J/cm^2$) pulses, which consists of 4 overlapping spots is presented. At the top of Fig.(5.30), the rotation angle exhibited $\alpha = 0°$. A ripple pattern oriented parallel as well as secondary structures oriented perpendicular to the polarization can be seen (indicated arrows).

At $\alpha = 30°$, the ripple orientation followed the WL polarization as expected. On the direct border of both spots ripples with both orientation can be found (circles). This mainly results from the beam filamentation of the WL, where pre-existing patterns could remain in non-affected areas. At the border between the $\alpha = 30°$ and $\alpha = 60°$ spots the same feature can be found. In this region, some lines from the secondary pattern of the lower $\alpha = 60°$ spot reach into the upper

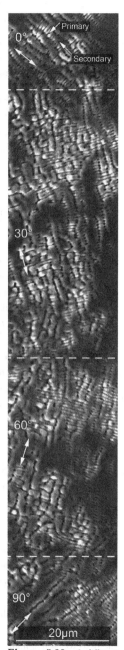

Figure 5.30: *3 different orient. of polarization.*

$\alpha = 30°$ spot. These results show that it is possible to control the LIPSS orientation with WL on a solid surface.

5.5 EDX-Measurements

In the last section of this chapter, *Energy Dispersive X-ray Spectroscopy* measurements on two ablation spot are presented. During the scanning the sample surface in the *SEM Zeiss EVO 40*, with a monochromatic electron beam, an X-ray spectrum resulting from the interaction of the primary electrons with the surface atoms, was recorded as well.

The X-ray spectrum consists of a continuous part and characteristic peaks. The incident electrons can be scattered by the coulomb field of an atomic nucleus, thereby giving up some or all of its energy. This energy may be emitted in the form of X-rays. Since the primary electron can give up any amount of its energy during this process, the energy distribution of the emitted X-rays is continuous.

The characteristic peaks of the X-ray spectrum arise from the interaction of the incident electron beam with electrons of an inner atomic shell. A bound electron is ejected, subsequently the vacancy is filled by an electron from an outer shell. The distinct energy difference between the shells is set free in the form of X-ray radiation, which are characteristic for each element [NIM99].

The nomenclature indicates the contributing shells, for instance, a transition from the L (n=2) shell to the lower K (n=1) shell is named K_α, and M (n=3) $\rightarrow$ K (n=1) is named K_β, correspondingly.

The source of X-ray radiation, in contrast to the secondary electrons from the direct surface layer, can come from up to $z = -3\mu$m depth below the surface, depending of the material and energy of the incident electron beam. The great advantage of EDX-method combined with SEM, is the possibility to map the distribution of the previously defined elements over the scanned area, while scanning the sample in a raster. Thereby a 2D image of the element composition on the surface can be generated.

One of the scanned samples is the ablation spot on n-silicon, from the beginning of the previous section (Fig.5.9), produced with 100, $E_d = 24.2\mu$J pulses, and with the @800nm filter. The SEM image of the mapped area is shown in Fig.(5.32). The sample was cleansed in an ultrasonic bath before the measurement, so no loose particles were left on the surface. The corresponding EDX spectrum of this area is displayed in Fig.(5.31).

In the spectrum, strong characteristic peaks of silicon, arsenic, oxygen and carbon can be found. The relative intensity of the element peaks is related to their occurrence on the surface of the sample. The plateau on the left-hand side as well as the decreasing part on the right-hand side, makes up the continuous part of the X-ray spectrum. The most intensive peak arises, of course, from silicon.

The silicon peak actually consists of three peaks, caused by the K_{α_1}, K_{α_2} and K_{β_1} radiation, at energies of 1.73998keV, 1.73938keV and 1.83594keV, respectively. However, the resolution of the detector is not high enough to display them as separate peaks. The corresponding X-ray energies are taken from the X-ray data booklet [Lab09].

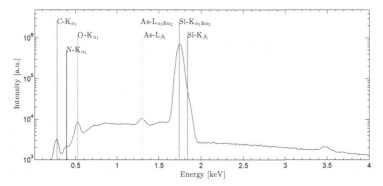

Figure 5.31: *EDX spectrum of mapped area presented in Fig.(5.32).*

The EDX analysis showed that the silicon sample is doped with arsenic. The corresponding $L_{\alpha_1 \& 2}$ and L_{β_1} peaks are located at energies of 1.282keV, 1.282keV and 1.317keV. At lower X-ray energies, the K_{α_1} peaks of carbon, Nitrogen and oxygen at 0.277keV, 0.392keV and 0.524keV can be found. The occurrence of these elements on the surface may arise from the fact, that the white-light ablation took place in air atmosphere. By taking a look at the distribution maps of the stronger peaks displayed in Fig.(5.33) this can be better understood. Only the strong characteristic peaks were mapped.

Figure (5.33(a)) shows the silicon distribution. As expected, it is evenly distributed, solely near the center, where the ablation spots is located, less silicon K_α X-ray radiation was detected, because of a possible absorption of the radiation in the surface layer.

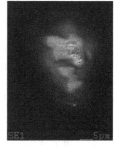

Figure 5.32: *SEM image of scanned EDX area on silicon. Region is compareable with Fig.(5.9(a)).*

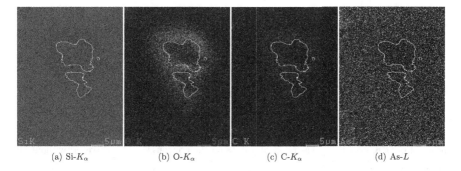

(a) Si-K_α (b) O-K_α (c) C-K_α (d) As-L

Figure 5.33: *X-ray K_α distribution maps of strong characteristic peaking elements, found on the surface of the silicon target after irradiation in air. Spot produced with 100, $E_d = 24.2\mu J$ pulses and with the @800nm filter. The white line shows the trace where LIPSS was produced.*

The oxygen map shows a very interesting feature (Fig.5.33(b)). It seems that the element oxygen is strongly located just around the area where LIPSS has formed (white trace). Inside the structured area, as well as on the rest of the sample, the oxygen appears evenly distributed. The same behavior can be found in the distribution map of carbon (Fig.5.33(c)). Around the trace of the LIPSS area, more carbon could be detected than on the rest of the sample. In the same region at the SEM image, the surface is modified yet no LIPSS formed on this outer region of the spot. This modified area around the structured part is often observed during femtosecond laser ablation of mono-crystalline silicon with oxide layer or ablation in air. In [BBM04] it has been reported, that in this region the crystalline state of the silicon changed to amorphous during irradiation.

The last X-ray distribution map in Fig.(5.33(d)) of the element arsenic, shows a slight decrease of the arsenic concentration inside the trace of the LIPSS structured area. This again can be explained by the fact that more matter was removed in this region than in the surrounding area. Additionally it is to say that a few more peaks of low intensity may be found in the X-ray spectrum. At lower energies they could arise from other trace elements in air. The peak at around $\approx 3.49\text{keV}$ may result from the L_{β_1} radiation of indium (3.487keV) as a possible co-dopand

The second spot investigated, was also produced on a silicon sample, under irradiation conditions of $N = 300000$ applied pulses at $E_d = 18.8\mu\text{J}$ and @800nm cut-off filter. This time that sample was not cleansed after irradiation.

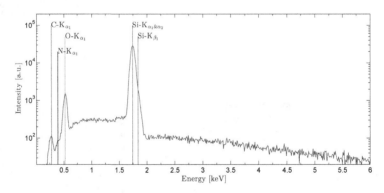

Figure 5.34: *EDX X-ray spectrum of mapped area presented in Fig.(5.35(a)).*

The corresponding X-ray spectrum shows in general the same characteristic $K-$radiation peaks of carbon, nitrogen, oxygen and silicon, as on the previous sample. However, without an arsenic dopand peak.

On the SEM image in Fig.(5.35(a)) no organized structures are visible, instead the spot is covered, of what appears as little grains. This feature has often been observed after long irradiation with low WL pulse energy, in air.

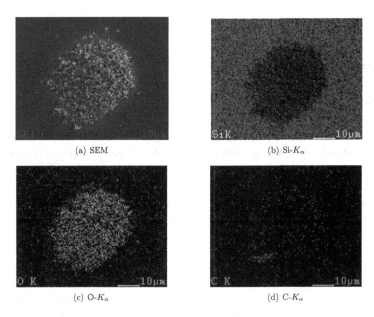

(a) SEM   (b) Si-K_α

(c) O-K_α (d) C-K_α

Figure 5.35: *Distribution maps of strong characteristic peaking elements, found on the ablation spot on the surface of the silicon target (@800nm filter, $N = 300000\#$, $E_d = 18.8\mu J$).*

The Si-K radiation map (Fig.5.35(b)) shows, that the silicon concentration is decreased at the site of the crater, whereas the on the oxygen map, the concentration of oxygen is heavily increased at the exact same position of the crater (Fig.5.35(c)). The corresponding carbon map does not show any significant increase in concentration, except for a small spot at the lower left side of the image, which is probably a piece of dirt on the sample (Fig.5.35(d)). This piece can also be identified at the SEM image.

From these distribution maps it seems, that the grains covering the crater are the reaction product SiO_X, of a thermal reaction of ejected silicon ions and atmospheric oxygen gas. However, from this analysis it cannot be clarified whether the product is SiO_1 or SiO_2, yet the latter is more likely. The growth of these SiO_X grains can be seen as well in Fig.(1.3) in the introduction part in chapter 1.2.

This analysis showed, that the silicon sample, heavily reacted with the oxygen gas during the irradiation of the sample in air. Especially in the non-structured areas a higher oxygen and carbon concentration was found.

Chapter 6

Conclusion & Outlook

This thesis combined the two major scientific fields, of *laser induced periodic surface structures* and *white-light supercontinuum generation* upon femtosecond laser irradiation. The aim was to generate ultrafast white-light pulses of low spatial coherence and use it to study the importance of wavelength and coherence vs. irradiation dose on the characteristic features of LIPSS.

The generated white-light continuum exhibited a broad continuous spectrum, with usable bandwidths of $\Delta\lambda_{wl}(@800) \approx 400\text{nm}$, $\Delta\lambda_{wl}(@700) \approx 300\text{nm}$ and sufficient white-light pulse energy for ablation. Upon small input pump pulse energies it was found, that the continuum still preserves the high spatial coherence of the pump-laser, noticeable in the interference on the screen, of the self-trapping filaments inside the crystal. However, at large input energy the interference patterns are averaged out, and a significantly reduced spatial coherence could be measured. At this point, the near-field intensity distribution revealed that the beam broke up in very many filaments along its transverse axis. This observation supports the thesis suggested in [BIC96], that the beam is a superposition of light emitted from many sources (filaments), with a large distribution of phases, i.e. the white-light is not spatially coherent.

During the experimental study of LIPSS, it could be demonstrated successfully, that periodic patterns form on the spot areas, upon irradiation with this continuous exciting spectrum of low spatial coherence. A coexistence of fine and coarse ripples as well as an evolution of the surface patterns upon multi-pulse irradiation was also observed.

Material:	Silicon	S. Steel	Copper	Brass
Periods:	590-750nm	290-420nm	300-520nm	$\approx 500\text{nm}$

Table 6.1: *Spatial periods of the primary ripple pattern after multi-pulse irradiation observed in this thesis.*

The spatial periods of the primary ripple patterns (overview in table 6.1), determined via FFT of the corresponding SEM images, are found to be insensitive to the wavelengths of the incident light. Even if the intensity peak in the corresponding spectra, is made responsible to excite surface plasmon polaritons and, thus, "lithographically imprint" the structures, no correlation between peak wavelength and smaller measured period lengths ($\Lambda_{pri} < \lambda_{wl,peak}$) can be identified. Instead, the measured period lengths exhibit a clear dependence on the individual material as well as on the applied irradiation dose. Furthermore, the highly developed pattern on the copper surface upon irradiation with a flatter spectrum (moderate slope in narrow spectral intervals), supports this picture.

The observed coalescence of the primary ripple pattern into a coarser secondary pattern and more irregular structures appears to be strongly correlated to the positive feedback and further depends on the local white-light intensity. On silicon the secondary patterns exhibited at least a period doubling compared to the spatial periods of the primary patterns. This is in agreement to previous reported observations [Var13b].

$$\Lambda_{\text{pri}} \propto \frac{1}{F} \cdot \exp\{-I_D\} \tag{6.1}$$

An exponentially decreasing behavior (Eq. 6.1) of the primary ripple periods, predicted by the *dynamic* model for the linear regime, could be found in the multi-pulse irradiation experiments, for the silicon and stainless steel samples. For copper and Brass the periods also decreased, however, more data will have to be gathered to identify the clear exponential behavior.

Being the key property of the incident light, it seems that no spatial coherence of the white-light / light source is required for the structures formation and ordering into period patterns on solid targets. The observed and investigated periodic surface structures exhibited the same common features like bifurcations or truncations as well as columnar, and bubble-like structures like periodic structures resulting from induced instability (e.g. ion beam sputtering, ripples in déserts and seashores). The spatial periods ranges equal the ranges of ripples produced under $\lambda = 800$nm laser irradiation conditions.

In conclusion, it seems unlikely to attribute the formation of periodic surface structures to an optical interference effect, given the continuous exciting spectrum with significantly reduced spatial coherence. Instead, it appears that the effect of structure formation depends, preferably on the deposited amount of energy. Therefore, an association of the observed pattern features to self-organization as the driving process in periodic structure formation appears justified.

Because of the complexity of both fields, a variety of attractive investigations is left open for future experiments. For once, the determination of the white-light pulse duration, after the propagation through the 5mm short Al_2O_3 crystal. One possible experiment can be done, by using the frequency-up conversion technique and overlap the pump-laser and the white-light continuum in an optical material of high non-linearity.

By using a monochromator the absolute power (W/nm) in the narrow spectral intervals may be obtained as well. The effect of self-focusing on the beam diameter could be investigated by determining the actual size of the self trapping filaments. Therefore, the beam imaging experiment, in section 4.4, can be further developed to image the focus of the beam inside the crystal.

As a result of the pump-laser limitation in pulse energy, the structuring of silicon with the @700nm cut-off spectrum was not possible. Dielectric materials were not dealt with, due to their high ablation threshold. Thus, if higher pump-laser energies are available, it would be of great interest to also produce structures for these materials and compare their characteristic features to the ones presented in here.

Moreover, the periodic patterns seem to only exist in a narrow interval of applied irradiation dose. For this reason, the irradiation dose ought to be changed, by varying the fluence and number of pulses. This may be accomplished by carefully adjusting the telescope arrangement, thereby adjusting the focus area, without lowering the white-light pulse energy, thus, lowering spectral bandwidth.

To further establish clarity of the ripple period insensitivity to the incident light wavelength and support the thesis in this work, several different experimental approaches may be performed. It is known that the spectral characteristics (spectral broadening & spectrum flatness (slope)) of the white-light continua, depend on the pump-laser wavelengths and bandgap of the generating medium. Therefore, the formation of the primary ripple pattern may be investigated with different white-light continua of different spectral characteristics.

Overall, more data will need to be gathered in the investigation of the spatial ripple periods upon multi-pulse irradiation experiments (produce structures with smaller pulse intervals), to improve the experimental statistics and reproducibility.

Appendix

A Temporal Intensity distribution and Pulse duration of the Laser-Beam

The oscillator (*Spectra Physics: Tsunami*), which generates the femtosecond laser pulsed is specified and setup in a way, to produce longitudinal resonator modes that are close to either a Gaussian Bell shape intensity distribution[1]:

$$I_{\text{Gauss}}(t) = I_0 \cdot \exp\left\{ -\frac{t^2}{\tau_1^2} \right\} , \tag{6.2}$$

or a hyperbolic secant (sech^2) shaped intensity distribution:

$$I_{\text{sech}^2}(t) = I_0 \cdot \text{sech}^2\left(\frac{t}{\tau_2}\right) = I_0 \cdot \frac{1}{\cosh^2(t/\tau_2)} , \tag{6.3}$$

or somewhat in between. The pulse duration was determined with the multi-shot autocorrelator (*Spectra Physics, Model 409*), which measured an intensity autocorrelation width at the FWHM of $\Delta\tau_A$ =106fs. The autocorrelation trace as well as the applied fits for Gaussian and Sech^2 is displayed in (A1)

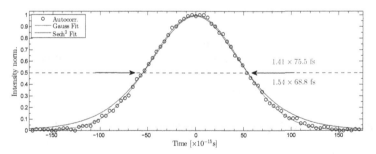

Figure A1: *Autocorrelation trace (circles) and Gaussian and Sech^2 fit.*

The value for the pulse duration τ_p, then is obtained over $\tau_p = \Delta\tau_{ac}/k$, where k is the deconvolution factor for the corresponding distribution. The autocorrelation width is in those two cases larger than the actual FWHM of the intensity distribution.

For a Gaussian intensity envelope k =1.41, while for Sech^2 k =1.54. This results in pulse durations of τ_p=75.5fs for a Gaussian pulse and τ_p=68.8fs for a Sech^2 pulse shape.

---

[1] The parameters $\tau_{1,2}$ are not to be mistaken for the pulse duration τ_p

From the corresponding fits (Fig.A1) of the autocorrelation data, one can presume that the laser pulse exhibits a Gaussian intensity distribution, rather than Sech². In order to further verify this result the *time bandwidth product* (TBP) can be calculated. The TBP is defined as:

$$TBP = \Delta \nu \cdot \tau_p , \qquad (6.4)$$

where $\Delta \nu$ is the spectral frequency width of the pulse at FWHM. For ideal Gaussian pulse envelopes the TBP is TBP_{Gauss}=0.441 and TBP_{sech}=0.315 for sech² envelopes. The quantity $\Delta \nu$ can be calculated from the total differential of the central laser frequency ν.

$$\Delta \nu = \left| \frac{\partial \nu}{\partial \lambda} \Delta \lambda \right| \qquad (6.5)$$

The spectral width in the wavelength domain of the laser was measured with a spectrometer to be $\Delta \lambda$=15nm. With the spectral width and the two pulse duration limits, the corresponding TBP's are TBP_{Gauss}=0.526 and TBP_{sech^2}=0.482. It seems obvious that the Gaussian time bandwidth product is closer to the fundamental value, which is a good evidence, that the laser pulse is likelier to exhibits a Gaussian intensity distribution with τ_p=75.2fs duration in the temporal domain, rather than sech² shaped envelope [Sch12]. The close proximity of the TBP to the transform-limited ideal case, indicates a good quality of the temporal pulse shape.

The laser pulses from the oscillator are subsequently amplified (*Spectra Physics: Spitfire*) and guided over 4 silver coated mirrors, a polarizer, a $\lambda/2$ wave-plate and a glass lens to the Al_2O_3 crystal (Fig.3.2). Along this way, the duration of the pulse may slightly increase, because of a positive group velocity dispersion caused especially by the propagation through the wave-plate and glass lens. For this reason, it was ensured that the pulse is shortest ($\tau_p \approx$ 75fs) at the position of the crystal, hence, the pulse of highest intensity generates the continuum.

This could be done by giving the output pulse from the amplifier a negative group velocity delay (through the translation stage of the pulse compression, subsequent to the *chirped amplification* arrangement), and measure the resulting power of the generated continuum. Thus, the shortest pulse generates the highest power continuum.

B Spatial Intensity Distribution and Propagation of the Laser-Beam

The Gaussian intensity distribution in the spatial domain as well as the beam propagation of the laser beam was determined with the monochrome (*EHD-12V5HC*) CCD-camera. The beam was focused with a f=+300mm plano-convex lens and CCD-images were taken across the waist area at different positions in Z-direction (direction of propagation) of the focused beam. The intensity of the focused beam was attenuated to the minimum and subsequently dimmed with 3 neutral density filters (NDF).

With a +300mm lens, the intensity in the focus could be better attenuated by the help of NDF, due to a wider waist and confocal parameter (double Rayleigh-range $b = 2 \cdot Z_R$) of the beam, compared to a +150mm lens used for continuum generation.

The measurements where taken around the position of the experimental setup of continuum generation. Out of each CCD-image the $1/e^2$ intensity widths $w_{x,y}$ in the transverse x&y-directions of the beam were measured. The data of the waist are plotted in Fig. (B2).

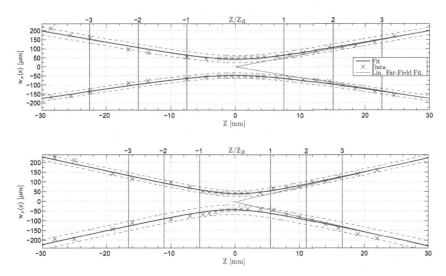

Figure B2: *Propagation of the output laser beam. (upper): Beam width in x-direction across the focus. (lower): width in y-direction across the focus. Crosses correspond to data, blue-solid-line represents Gaussian beam fit. Dashed-line shows 1σ standard deviation. The upper x-axes are scaled to the Rayleigh range.*

In the upper plot, the propagation of the horizontal beam width w_x (with respect to to the worktable) and in the lower plot the corresponding vertical beam width w_y is displayed . The data points are fitted with the Eq.(6.6), which is the formula for Gaussian $1/e^2$-intensity width in direction of propagation (Z-direction).

$$w\left(Z\right) = w(M^2)_{0,300} \cdot \sqrt{1 + \left(\frac{Z}{Z_{R,300}}\right)^2} . \qquad (6.6)$$

From the fit the radius of the focused beam at the waist $w_{0,300}$, as well as the value for the Rayleigh range $Z_{R,300}$ is obtained. A complete evaluation of the individual widths is given in table (6.2). The quality of the beam is determined over the beam parameter product defined as:

$$w_0 \cdot \theta = M^2 \cdot \frac{\lambda}{\pi} , \qquad (6.7)$$

where θ is the divergence angle of the beam. The M^2 factor is related to the beam quality factor K, given as $K = 1/M^2$. For an ideal Gaussian shaped beam the beam parameter product can reach the lower fundamental limit of $w_0 \cdot \theta = \lambda/\pi$ with an M^2=1, thus, $K = 1$. However, in the real case the M^2 factor is always $M^2 > 1$ an therefore a beam quality factor $K < 1$.

The divergence angle θ is obtained by fitting $w_{x,y}$ at least $3 \times Z_{R,300}$ Rayleigh-ranges outside the waist w_0. Here, $w_{x,y}$ is supposed to increase linearly with distance Z [EEi10]. The linear fit is represented by the red-solid line in Fig.(B2). The slope then equals the divergence angle θ in radiant units. With that, the M^2 factor and the beam quality factor K can be calculated (displayed in table (6.2)).

	Direct.	$w_{0,300}$ [μm]	$\theta \times 10^{-5}$ [rad]	M^2	K=1/M^2	K [%]
	x	42.41 $\pm$ 3.4	663 $\pm$ 0.92	1.102 $\pm$ 0.09	0.907 $\pm$ 0.07	
	-x	47.81 $\pm$ 2.4	588 $\pm$ 1.69	1.104 $\pm$ 0.05	0.906 $\pm$ 0.05	
	y	38.81 $\pm$ 2.7	761 $\pm$ 0.90	1.159 $\pm$ 0.08	0.863 $\pm$ 0.06	
	-y	43.18 $\pm$ 4.1	761 $\pm$ 1.28	1.290 $\pm$ 0.12	0.775 $\pm$ 0.07	
Σ_x	X	45.11 $\pm$ 2.9	626 $\pm$ 1.31	1.103 $\pm$ 0.07	0.907 $\pm$ 0.06	91 $\pm$ 6%
Σ_y	Y	40.99 $\pm$ 3.4	761 $\pm$ 1.09	1.225 $\pm$ 0.10	0.819 $\pm$ 0.07	82 $\pm$ 7%

Table 6.2: *Fit coefficients and errors. Presented are beam waist sizes $w_{0,300}$, beam divergence angle θ, M^2 factor and beam quality factor K, the $\pm$ corresponds to the calculated errors. $\Sigma_{x,y}$ is an average for x&y-direction. The last 2 rows (Σ_i) are the average values of the two x and y values. The errors for $w_{0,300}$ and θ were obtained from the fit, while the M^2 and K errors were calculated though error propagation.*

The measured widths show, that divergence of the beam is greater in y-direction (vertical), than in the y-direction (horizontal). With an M^2 factor of M_x^2=1.103 and M_y^2=1.225 the output laser beam exhibits a good spatial profile, which is close to an ideal Gaussian beam. This means, that the output beam is diffraction limited with a beam quality is 91% in the horizontal and 82% in the vertical direction.

Through the measurement of the beam propagation, the position of the waist could also be found. In the x-direction the average width (radius) at the waist is $\bar{w}_{0x,300}$=45.11μm, hence the beam has a $\bar{d}_{0x,300} = 90.22\mu$m diameter in the focus. Correspondingly, in y-direction the waist size is $\bar{w}_{0y,300}$=40.99μm and the diameter $\bar{d}_{0y,300} = 81.98\mu$m.

In order to retrieve the waist of the +150mm lens one uses Eq.(6.8).

$$w_{0,f} = \frac{\lambda\, f}{\pi\, w_0} \qquad (6.8)$$

This equation applies, if the position of the waist coincides with the lenses focal length f, experimentally this is generally the case if the beams initial Rayleigh-range of the unfocused beam exceeds the focal length by far $Z_R \gg f$. The parameter w_0 is theoretically the waist size of the unfocused beam . By rearranging Eq.(6.8) and equating with the quantities for the f_1=+300mm and f_2=+150mm lenses, Eq.(6.9) can be obtained:

$$w_{0,150} = \frac{f_2}{f_1} \cdot w_{0,300} \ . \qquad (6.9)$$

The same applies for the divergence angles θ_1 (+300mm lens) and θ_2 (+150mm lens), however, the ratio of the focal lengths is inverted, since the divergence angle becomes larger at stronger focus, due to diffraction. With the constant beam parameter product Eq.(6.7) and Eq.(6.9), one gets:

$$\theta_2 = \frac{f_1}{f_2} \cdot \theta_1 \ . \tag{6.10}$$

The calculated waist sizes and divergence angles of the beam focused with the +150mm lens are given in table 6.3. This equals a beam area of $A_{0,150} \approx 1452\mu m^2$ in the focus of the +150mm lens, and for the maximum input pulse energy of $E_0 \approx 270\mu J$ an intensity of $I \approx 248$ TW/cm^2 (calculated through Eq. 3.1). This intensity value is considerably increased inside the Al$_2$O$_3$ crystal due to self-focusing.

Figure (B3) shows a 3 dimensional visualization of one Gaussian $\tau_p = 75$fs laser pulse at the waist of the beam (in the focus of the +300mm lens) with the measured parameters from above.

	$w_{0,150}$ [μm]	$\theta \times 10^{-5}$ [rad]
X	22.55 ± 1.45	1252 ± 1.62
Y	20.49 ± 1.70	1522 ± 2.18

Table 6.3: *Calculated waist sizes (radius at waist) in transverse directions and divergence angle of the laser beam if focused with a f = +150mm lens. Errors calculated though error propagation.*

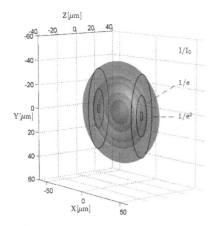 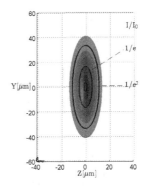

Figure B3: *Intensity distribution of one 75fs ($\lambda = 800nm$) laser pulse at the waist of the beam in the +300mm lens focus. False color intensity layers represent values of $I/I_0 = 0.135, 0.367, 0.5, 0.75, 0.9$. The three contour lines represent values of $I/I_0 = 0.135, 0.367, 0.5$ at $X = \pm 30\mu m$ from the center ($X_c = 0\mu m$). The $\tau = 75fs$ pulse duration at the FWHM has been recalculated to the $1/e^2$ value ($\tau_{e^{-2}} \approx 90.1fs$, full width @$e^{-2}$) and transformed to a length via $c \cdot \tau_{e^{-2}} = x_\tau \approx 27\mu m$, with c being the speed of light.*

The Intensity distribution is calculated with equation (6.11), where $w_x(z), w_y(z)$ are obtained from Eq. (6.6):

$$I(x,y,z) = I_0 \cdot \left(\frac{w_{0x,300}}{w_x(z)} \cdot \frac{w_{0y,300}}{w_y(z)} \right)^2 \cdot \exp\left(-2 \cdot \left(\frac{x^2}{w_{0x,300}^2} + \frac{y^2}{w_{0y,300}^2} + \frac{z^2}{x_\tau^2} \right) \right) \ . \tag{6.11}$$

Bibliography

[Alf06]   Robert R. Alfano. *The Supercontinuum Laser Source - Fundamentals with Updated References.* Springer, Berlin, Heidelberg, 2nd edition, 2006.

[ASh70]   R. Alfano and S. Shapiro. Observation of self-phase modulation and small-scale filaments in crystals and glasses. *Physical Review Letters*, 24(11):592–594, 1970.

[BBM04]   J. Bonse, K. Brzezinka, and A. Meixner. Modifying single-crystalline silicon by femtosecond laser pulses: an analysis by micro raman spectroscopy, scanning laser microscopy and atomic force microscopy. *Applied Surface Science*, 221(14):215 – 230, 2004.

[BCh99]   A. Brodeur and S. L. Chin. Ultrafast white-light continuum generation and self-focusing in transparent condensed media. *Optical Society of America*, 16(4):637–650, 1999.

[BHa88]   M. Bradley and J. Harper. Theory of ripple topography induced by ion bombardment. *Journal of Vacuum Science and Technology A*, 6(4):2390–2395, 1988.

[BHä00]   M. Bellini and T. Hänsch. Phase-locked white-light continuum pulses: toward a universal optical frequency-comb synthesizer. *Opt. Lett.*, 25(14):1049–1051, 2000.

[BIC96]   A. Brodeur, F. Ilkov, and S. Chin. Beam filamentation and the white light continuum divergence. *Optics Communications*, 129(34):193 – 198, 1996.

[Bir65]   Milton Birnbaum. Semiconductor surface damage produced by ruby lasers. *Journal of Applied Physics*, 36:3688–3689, 1965.

[Boa82]   A.D. Boardman. *Electromagnetic Surface Modes.* Wiley & Sons Ltd., 1982.

[Boy03]   Robert W. Boyd. *Nonlinear Optics.* Academic Press, 2nd edition edition, 2003.

[BRK09]   J. Bonse, A. Rosenfeld, and J. Krüger. On the role of surface plasmon polaritons in the formation of laser-induced periodic surface structures upon irradiation of silicon by femtosecond-laser pulses. *Journal of applied physics*, 106:104910 1–5, 2009.

[CHR03]   F. Costache, M. Henyk, and J. Reif. Surface patterning on insulators upon femtosecond laser ablation. *Applied Surface Science*, 208-209:486–491, 2003.

[Com14]   Laser Components. Interexchange with Björn Götze, February 2014. b.goetze@lasercomponents.com.

[Cos06]   Florena Costache. *Dynamics of Ultra-short Laser Pulse Interaction with Solids at the Origin of Nanoscale Surface Modification.* PhD thesis, Brandenburg University of Technology Cottbus, 2006.

[CPB+99]   S. Chin, S. Petit, F. Borne, et al. The white light supercontinuum is indeed an ultrafast white light laser. *Japanese Journal of Applied Physics*, 38(2A):L126, 1999.

[CPe92] P. Chernev and V. Petrov. Self-focusing of light pulses in the presence of normal group-velocity dispersion. *Opt. Lett.*, 17(3):172–174, 1992.

[CTu30] M. Czerny and A. Turner. Über den Astigmatismus bei Spektrometern. *Z. Physik*, 61:792, 1930.

[CVI14] Laser Corporation CVI. Interexchange, February 2014. www.cvilaseroptics.com.

[Dem13] Wolfgang Demtröder. *Elektrizität und Optik : mit 19 Tabellen, zahlreichen durchgerechneten Beispielen und 145 Übungsaufgaben mit ausführlichen Lösungen.* Springer, Berlin, 2013.

[DRM06] A. K. Dharmadhikari, F. A. Rajgara, and D. Mathur. Depolarization of white light generated by ultrashort laser pulses in optical media. *Opt. Lett.*, 31(14):2184–2186, 2006.

[EEi10] H. Eichler and J. Eichler. *Laser.* Springer, 2010.

[Ekv00] Katrin Ekvall. *Time Resolved Laser Spectroscopy: Non-linear polarisation studies in condensed phase andLifetime studies of alkaline earth hydrides.* PhD thesis, Royal Institute of Technology Stockholm, 2000.

[GCP+11] F. Garrelie, J. Colombier, F. Pigeon, et al. Evidence of surface plasmon resonance in ultrafast laser-induced ripples. *Optics Express 9035*, 19(10), 2011.

[Gol90] I. Golub. Optical characteristics of supercontinuum generation. *Opt. Lett.*, 15(6):305–307, 1990.

[GSE86] I. Golub, R. Shuker, and G. Erez. On the optical characterization of the conical emission. *Optics Communications*, 57(2):143–145, 1986.

[HKF+04] R. Holenstein, S. E. Kirkwood, R. Fedosejevs, et al. Simulation of femtosecond laser ablation of silicon. *Photonic Applications in Telecommunications, Sensors, Software, and Lasers*, 688, 2004. Photonics North 2004 Conference Volume 5579.

[HLC+01] I. Hartl, X. D. Li, C. Chudoba, et al. Ultrahigh-resolution optical coherence tomography using continuum generation in an air–silica microstructure optical fiber. *Opt. Lett.*, 26(9):608–610, 2001.

[ImF12] T. Imran and G. Figueira. Intensityphase characterization of white-light continuum generated in sapphire by 280 fs laser pulses at 1053 nm. *Journal of Optics*, 14(3):035201, 2012.

[Inc14] Thorlabs Inc. FESH0700 premium short pass filter, May 2014. www.thorlabs.de/images/popupimages/FESH0700.xlsx.

[JGL+02] H. Jeschke, M. Garcia, M. Lenzner, J. Bonse, J. Krüger, and W.Kautek. Laser ablation theshold of silicon for different pulse durations: theory and experiment. *Applied Surface Science*, 197-198:839–844, 2002.

[JSM+05] C. Jarman, D. Schumacher, C. Modoran, et al. Supercontinuum generation in sapphire: A measurement of intensity. Technical report, Ohio State University, 2005.

[KSE⁺00]   D. Kip, M. Soljacic, M. Segevand E. Eugenieva, et al. Modulation instability and pattern formation in spatially incoherent light beams. *Science*, 290(5491):495–498, 2000.

[KTs76]   Y. Kuramoto and T. Tsuzuki. Persistent propagation of concentration waves in dissipative media far from thermal equilibrium. *Progress of Theoretical Physics*, 55(2):356–369, 1976.

[Lab09]   Lawrence Berkeley National Laboratory. X-ray data booklet, 2009. Center for X-ray Optics and Advanced Light Source, xdb.lbl.gov.

[LLM03]   P. Lorazo, L. Lewis, and M. Meunier. Short-pulse laser ablation of solids: From phase explosion to fragmentation. *Physical Review Letters*, 91(22), 2003.

[LLS⁺05]   A. Lindenberg, J. Larsson, K. Sokolowski-Tinten, et al. Atomic scale visualization of intertial dynamics. *Science*, 308, 2005.

[LWM⁺94]   G. Luther, E. Wright, J. Moloney, et al. Self-focusing threshold in normally dispersive media. *Opt. Lett.*, 19(12):862–864, 1994.

[Mai60]   T. Maiman. Stimulated optical radiation in ruby. *Nature*, 187(187):493–494, 1960.

[Mar75]   J Marburger. Self focusing theory. *Prog. Quantum Electron*, 4(35), 1975.

[MCN⁺96]   C. Momma, B. Chichkov, S. Nolte, et al. Short-pulse laser ablation of solid targets. *Optics communications*, 129(1):134–142, 1996.

[MKe95]   A. Miotello and Roger Kelly. Critical assessment of thermal models for laser sputtering at high fluneces. *Applied Physics Letters*, 67(24), 1995.

[MPo00]   V. Margetic, A. Pakulev, and A. Stockhaus others. A comparison of nanosecond and femtosecond laser-induced plasma spectroscopy of brass samples. *Spectrochimica Acta Part B*, 55:1771–1775, 2000.

[NIM99]   Wisconsin NORAN Instruments Middleton. Energy - dispersive x-ray microanalysis - an introduction, 1999. , www.noran.com.

[NOS⁺96]   H. Nishioka, W. Odajima, Y. Sasaki, et al. Super-wide-band coherent light generation in multi-channeling propagation of tera-watt ti: Sapphire laser pulses. *Progress in Crystal Growth and Characterization of Materials*, 33(13):237 – 240, 1996.

[NSo10]   J. Nielsen, J. Savolainen, and M. S. Christensen others. Ultra-short pulse laser ablation of metals: threshold fluence, incubation coefficient and ablation rates. *Applied Surface Science*, 101:97–101, 2010.

[PJ14]   Jason Pomerantz and Juampiter, January 2014. www.redbubble.com/people/jasonpomerantz/works/6357267-sand-ripples-namib-desert www.flickr.com/photos/juampiter/5849791701/.

[Qio14]   Qiotiq. +10mm lens specifications, July 2014. Conversation with Qioptiq technical manager sandra.lotz@qioptiq.de.

[QWW⁺01]   Y. Qin, D Wang, S. Wang, et al. Spectral and temporal properties of femtosecond white-light continuum generated in h20. *Chinese Physics Letters*, 18(3):390, 2001.

[RCV⁺07] J. Reif, F. Costache, O. Varlamova, et al. Femtosecond laser induced surface instability resulting in self-organized nanostructures. In *Proc. SPIE 6606*, pages
66060G–66060G–8, 2007.

[Rei10] Juergen Reif. *Laser-surface interactions for new materials production tailoring
structure and properties*, volume 130, chapter Basic Physics of Femtosecond Laser
Ablation, pages 19–41. Springer, Berlin Heidelberg, 2010.

[RQu10] L. Ran and S. Qu. Structure formation on the surface of alloys irradiated by
femtosecond laser pulses. *Applied Surface Science*, 256:2315–2318, 2010.

[RVC08] J. Reif, O. Varlamova, and F. Costache. Femtosecond laser induced nanostructure
formation: self-organization control parameters. *Applied Physics A*, 92(4):1019–
1024, 2008.

[Sah10] S.K. Saha. *Aperture Synthesis: Methods and Applications to Optical Astronomy*.
Astronomy and Astrophysics Library. Springer, 2010.

[Sch12] Apl. Prof. Dr. R. Schmid. Ultraschnelle Optik & Femtochemie. Vorlesung Brandenburg University of Technology Cottbus, Sommersemester 2012.

[Sch14] Schott AG. *Optisches Glas Datenblätter*, July 2014. www.schott.com/advanced_
optics/german/abbe_datasheets/schott_datasheet_all_german.pdf.

[SCo94] D. Strickland and P. Corkum. Resistance of short pulses to self-focusing. *J. Opt.
Soc. Am. B*, 11(3):492–497, 1994.

[SCS99] K. Sokolowski-Tinten, A. Cavalleri, and C Siders. Detection of nonthermal melting
by ultrafast x-ray diffraction. *Science*, 286:1340–1342, 1999.

[SGo64] A. Savitzky and M. Golay. Smoothing and differentiation of data by simplified least
squares procedures. *Analytical Chemistry*, 36(8):1627–1639, 1964.

[Sha06] K. K. Sharma. *Optics principles and applications*. Academic Press, Amsterdam
Boston, 2006.

[Sig69] Peter Sigmund. Theory of sputtering. i. sputtering yield of amorphous and poly-
crystalline targets. *Physical Review*, 184(2):383–416, 1969.

[SMo85] D. Strickland and G. Mourou. Compression of amplified chirped optical pulses.
Optics Communications, 56(3):219 – 221, 1985.

[Son14] Sony. *Diagonal 8mm (Type 1/2) CCD Image Sensor for CCIR Black-and-White
Video Cameras*, 2014.
pdf.datasheetcatalog.com/datasheet/sony/a6802316.pdf.

[SSt82] M. Soileau and E. Van Stryland. Ripple structures and enhanced absorption associated with ordered surface defects. In *Laser induced damage in optical materials*,
pages 406–414, 1982.

[SYP⁺83] J. Sipe, J. Young, J. Preston, et al. Laser-induced periodic surface structure i.
therory. *Physical Review B*, 27(2):1141–1154, 1983.

[TBS04] C. Tal, H. Buljan, and M. Segev. Spontaneous pattern formation in a cavity with
incoherent light. *Opt. Express*, 12(15):3481–3487, 2004.

[TIm09]   G. Figueira T. Imran. Efficient white-light continuum generation in transparent
            solid media using 250 fs, 1053 nm laser pulses. In *Light at Extreme Intensities -
            LEI09*, 2009.

[Tos14]   Toshiba. Linear image sensor tcd1201d, Feburary 2014.
            www.stellarnet.us/public/download/TCD1201D.pdf.

[TTB+04]   S. Tal, C. Tal, H. Buljan, et al. Spontaneous pattern formation with incoherent
            white light. *Phys. Rev. Lett.*, 93, 2004.

[Var13a]   Olga Varlamova. Influence of laser irradiation dose on period of self-organized pat-
            terns. Technical report, Brandenburg University of Technology Cottbus, November
            2013. Weihnachtstagung.

[Var13b]   Olga Varlamova. *Self-organized surface pattern originating from femtosecond laser-
            induced instability*. PhD thesis, Brandenburg University of Technology Cottbus,
            2013.

[VCR+05]   O. Varlamova, F. Costache, J. Reif, et al. Self-organized pattern formation upon
            femtosecond laser ablation by circularly polarized light. *Applied Surface Science*,
            252:4702–4706, 2005.

[VCR+07]   O. Varlamova, F. Costache, M. Ratzke, et al. Control paramters in fattern formation
            upon laser ablation. *Applied Surface Science*, 253:7932–7936, 2007.

[Wik14]   Wikipedia, May 2014. Picture adapted from
            en.wikipedia.org/wiki/Gaussian_beam.

[WIt01]   W. Watanabe and K. Itoh. Spatial coherence of supercontinuum emitted from
            multiple filaments. *Japanese Journal of Applied Physics*, 40(2R):592, 2001.

[WYa97]   K. Wilson and V. Yakovlev. Ultrafast rainbow: tunable ultrashort pulses from a
            solid-state kilohertz system. *J. Opt. Soc. Am. B*, 14(2):444–448, 1997.

[XYA93]   Qirong Xing, Kwong Mow Yoo, and Robert R. Alfano. Conical emission by
            four-photon parametric generation by using femtosecond laser pulses. *Appl. Opt.*,
            32(12):2087–2089, 1993.

[YLW00]   T. Yau, C. Lee, and J. Wang. Femtosecond self-focusing dynamics measured by
            three-dimensional phase-retrieval cross correlation. *J. Opt. Soc. Am. B*, 17(9):1626–
            1635, 2000.

[YSh84]   G. Yang and Y. Shen. Spectral broadening of ultrashort pulses in a nonlinear
            medium. *Opt. Lett.*, 9(11):510–512, 1984.

[ZCF+09]   B. Ziberi, M. Cornejo, F. Frost, et al. Highly ordered nanopatterns on Ge and
            Si surfaces by ion beam sputtering. *Journal of Physics: Condensed Matter*,
            21(22):224003, 2009.

Acknowledgments

I hereby would like to take the chance and express my deepest gratitude to the two supervisors of this work, Prof. Dr. Jürgen Reif and Prof. Dr. Reiner Schmid. Thank you for letting me work in this interesting scientific field of research. Thank you for always having a spare minute for fruitful discussions and conversations.

The positive outcome of this work would have not been possible without a great team of scientists. Dr. Olga Varlamova and Dr. Markus Ratzke, thank you for your support in theoretical and technical questions. Frau Marion Borrmann, Jürgen Bertram and the rest of the Experimental Physics II chair at BTU Cottbus-Senftenberg for any kind of help during the time of this theses. I also thank Dr. Michael Krause and Dr. Winfried Seifert from Joint Lab for the help with the SEM Micrographs. Dr. Tzanimir Argiurov, thank you for the fruitful discussions at the beginning of this work.

Also, I thank the Springer Verlag for the honor of this award.

Lastly, I would like to express my thanks to my wife and my family for their long and enduring support.

Printed in the United States
By Bookmasters